D1222217

Electrons in
Chemical Reactions:
First Principles

Electrons in Chemical Reactions: First Principles

LIONEL SALEM
CNRS
and
Université de Paris-Sud

1807 1982

A Wiley-Interscience Publication
JOHN WILEY & SONS

New York   Chichester   Brisbane   Toronto   Singapore

Copyright © 1982 by John Wiley & Sons, Inc.

All rights reserved. Published simultaneously in Canada.

Reproduction or translation of any part of this work
beyond that permitted by Section 107 or 108 of the
1976 United States Copyright Act without the permission
of the copyright owner is unlawful. Requests for
permission or further information should be addressed to
the Permissions Department, John Wiley & Sons, Inc.

Library of Congress Cataloging in Publication Data:

Salem, Lionel.
  Electrons in chemical reactions.

  "A Wiley-Interscience publication."
  Includes bibliographical references and index.
  1. Chemical reaction, conditions and laws of.
2. Electronic structure. I. Title.

QD501.S223   541.3'9   81-19833
ISBN 0-471-08474-3   AACR2

Printed in the United States of America

10 9 8 7 6 5 4 3 2 1

126022

BELMONT COLLEGE LIBRARY

QD
501
S223
1982

Preface

Research is discovery, but also full understanding. In this respect this book is a research book, since in many sections I have attempted to go further and deeper than the usual interpretation or explanation as it now stands in the chemical literature.

Some readers will find that I have been short on references. Indeed, I have purposely taken a historical perspective and tried to restrict the 1970s references to a share that was proportional to their actual contribution to advancing our understanding rather than to their volume. Hence there are a relatively large number of references from the 1950s and 1960s and even the 1930s and 1940s.

I am grateful to Professors Gerald Segal, James McIver, Weston Borden, Nguyen Trong Anh, Josef Michl, Ken Houk, Olivier Kahn, and Alberte Pullman for their constructive comments on the successive chapters of the book. Monique Grisez and Françoise Pariset have also been of outstanding assistance.

Finally, this book reflects a theoretician's viewpoint and is a theoretician's translation of the experimental chemist's language. It owes its existence to *all* the scientists who in the last half century have made theoretical chemistry a living subject.

LIONEL SALEM

Laboratoire de Chimie Théorique
Université de Paris-Sud
31405 Orsay, France
January 1982

Contents

6. REACTION PATHS FROM THE PROPERTIES OF REACTANTS   158

7. THE ROLE OF SPIN: THE VARIOUS MANIFESTIONS OF ELECTRON EXCHANGE   189

8. SOLVENT PROPERTIES AND THE DIFFERENT MODELS FOR STUDYING SOLVENT EFFECTS   214

Electrons in
Chemical Reactions:
First Principles

Introduction:
Resonance in Chemistry

Resonance, introduced by Pauling and Wheland,[1] is the conceptual heart of chemistry. The notion that a molecule "hesitates" between different structures, borrows its characteristics from all of these and finally adopts a structure that is somewhat intermediate between them, is central to understanding electronic behavior.*

Resonance may involve alternative patterns for the distribution of double and single bonds, as in the famous case of benzene:

It may involve competition between convalent and ionic character, as in hydrogen chloride or in a phosphonium ylide:

$$H-Cl \leftrightarrow \overset{\oplus}{H}\overset{\ominus}{Cl} \qquad \overset{\ominus}{R_2C} - \overset{\oplus}{PR_3} \leftrightarrow R_2C{=}PR_3$$

It may illustrate the manner in which a net charge is actually distributed over several atoms, as in the enolate anion, in which case redistribution of unsaturation also occurs. Multiple possibilities may exist in the presence of unpaired electrons such as in ozone, where an unpaired diradical structure and zwitterionic structures are simultaneously available:

(see Section 3.5).

*For a lucid account of the history of structural uncertainty in organic chemistry, see Ref. 2.

1

TABLE I-1
Isosymmetric Isoelectronic Table for Simple Groups[a]

σ	π		
	0	**1**	**2**
0	>C⊕—	>Ċ— >C=	>C̈⊖—
		>N⊕— >N⊕=	>N̈—
			>Ö⊕—
1	>Ċ⊕—	>Ċ— >C=	>C̈⊖—
		>N⊕— >N⊕=	>N̈—
			>Ö⊕—
2	>C̈—	>Ċ⊖— >C̈⊖=	>N̈⊖—
	>N̈⊕—	>N̈— >N̈=	>Ö—
		>Ö⊕— >Ö⊕=	
3			⊖Ö̈—

[a] The number of electrons refers to the *labile electrons* at the atom (each σ bond adds nonlabile σ electrons in the immediate vicinity of the central atom). Dots *above* the atom refer to π electrons, dots *below* the atom refer to σ electrons. Net charge increases by 1 in each column from left to right and from second to third and fourth lines.

Electron counting is a crucial requirement in understanding the chemical behavior of molecules. For the rare reader who might not yet be familiar with this process, we give the electron count for some fundamental entities (Table I-1).

In this "isosymmetric, isoelectronic table" we have aligned groups that have not only the same number of total electrons, but also the same number of *both* σ and π electrons. By σ or π we mean to identify the nature of the atomic, hybrid, or molecular orbital occupied by each electron and its *symmetry*—σ, symmetric, or π antisymmetric—relative to the local plane of the bonds surrounding the atom (for an atom with a single bond to it, we proceed by continuity). It is clear that these symmetries serve

1. To distinguish electronic structures that have the same global count.

2. To refine the isoelectronic concept since chemical analogies will certainly be greater between two groups centered on different atoms and involving the same total number of electrons—but with *same* σ and π numbers—than between two groups with the same central atom but with *different* σ and π numbers.

3. Eventually, to develop rules for reactions between groups, based on the changes in σ and π counts during the reaction.

In Sections 1.2 and 1.3 the reader will find how resonance structures can be translated into a more quantitative language. In Chapter 3 a detailed study of the interplay of molecular orbital theory, valence–bond theory and resonance theory is given for a number of intermediates.

References

1. L. Pauling and G. W. Wheland, *J. Chem. Phys.* **1**, 362 (1933); L. Pauling, *The Nature of the Chemical Bond* (Cornell University Press, Ithaca, NY, 1960), p. 220.
2. C. A. Russell, *The History of Valence* (Leicester University Press, Leicester, UK, 1971), pp. 296 sqq.

1

Methods and Methodology

In the last half century (1930–1980) chemists have hardly modified their conceptual thinking on electrons in molecules. Either the electrons are regarded as being paired in bonds, according to the principles introduced by G. N. Lewis, or they are considered to exist as delocalized entities covering the entire molecule. In the first case the appropriate description of the electron pair is a *valence-bond* wave function, and in the second case each electron is described by a *molecular orbital* wave function. The actual numerical methods for calculating energies and chemical properties are far more numerous in the molecular orbital realm, although interest in valence-bond methods is increasing.

1.1 Wave Functions and Electronic States

Practically all studies of organic and inorganic reactions to date have been conducted within the framework of the Born-Oppenheimer approximation.[1] It is recognized that electrons are much lighter than nuclei (by a factor of at least 1836, the mass of the proton) and thereby move much faster. In the time it takes the electrons of a molecule to explore the entire space around all the nuclei, these have essentially remained at standstill. It is then common practice, except in sophisticated studies of nuclear-electronic coupling in the reactions of light molecules, to calculate all the electronic properties of molecules as if the nuclei were fixed. Hence, for *each* molecular geometry, the nuclei are given fixed positions, for which electronic wave functions and electronic stationary states are calculated. There will be a ground electronic state, with energy E_0 (or simply E) and excited electronic states, with energies E_1, E_2, E_3, and so on.

These electronic energies are obtained from the fixed nucleus wave equation

$$(V_{nn} + V_{ne} + V_{ee} + T_e)\,\Psi = E\Psi \tag{1-1}$$

where V_{nn}, V_{ne}, and V_{ee} describe the coulombic interactions between nuclei, between nuclei and electrons, and between electrons, respectively, and T_e represents the kinetic energy of the electrons.

If we now repeat the same operation for a different fixed-nucleus configuration, we get a new set of energies E_0', E_1', and so forth. For a given state, the set of energies E, E', E'', and so on form a *potential surface* (Fig. 1-1). Since each point on the surface corresponds to a different nuclear configuration, this surface actually describes the variation of molecular energy as a function of nuclear coordinates.

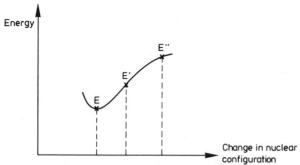

FIGURE 1-1. Potential surface for an electronic state.

The potential surface can then be understood as describing the potential energy to which the nuclei are subjected. In a dynamic manner the molecule can be regarded as moving *on* this surface.

1.2 Valence-Bond Wave Function for Two Electrons in a Covalent Bond

Heitler and London[2] introduced for the hydrogen molecule a function

$$\Psi^{cov} = \frac{1}{\sqrt{2 + 2S^2}} [\phi_A(1)\phi_B(2) + \phi_B(1)\phi_A(2)] \times \frac{1}{\sqrt{2}}(\alpha_1\beta_2 - \beta_1\alpha_2) \quad (1\text{-}2)$$

that has yet to be surpassed in its simplicity. It expresses the fact that when electron 1 is on atomic orbital ϕ_A, electron 2 is on atomic orbital ϕ_B, and conversely. In equation (1-2) S is the overlap integral

$$S = \int \phi_A\phi_B \, d\tau \quad (1\text{-}3)$$

between the two atomic orbitals ϕ_A and ϕ_B. The spin part of the wave function (α_1, upspin for electron 1; β_2, downspin for electron 2) shows the electrons to be paired in a singlet state. If the two electrons were present in the same bond but unpaired with parallel spins, the wave function would be that of the excited triplet state:

$$^3\Psi = \frac{1}{\sqrt{2 - 2S^2}} [\phi_A(1)\phi_B(2) - \phi_B(1)\phi_A(2)] \times \frac{1}{\sqrt{2}}(\alpha_1\beta_2 + \beta_1\alpha_2) \quad (1\text{-}4)$$

The corresponding potential curves drawn as a function of the internuclear distance R are shown in Fig. 1-2.

   The wave function for the ground singlet state has the following characteristics, which are typical of simple valence-bond functions: (1) the bonding is covalent, with the electrons either sharing the bond region or each occupying alone, at a given time, the region of an atom; (2) the electrons are relatively well "correlated,"

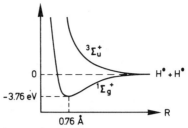

FIGURE 1-2. Potential curves for ground singlet state and lowest triplet state of H_2 using the Heitler-London wave function with strict atomic $1s$ orbitals of exponent unity. Reprinted from *Introduction to Quantum Mechanics* by L. Pauling and E. B. Wilson. Copyright, 1935, by the McGraw-Hill Book Company. Used with the permission of the McGraw-Hill Book Company.

keeping away from each other; and (3) the two electrons do not come together (except in the interatomic region) to profit simultaneously from the attraction of one of the nuclei. However, a consequential strength is the proper behavior of the function at large interatomic distances, where the molecule dissociates into two atoms.

The triplet wave function shares these features with the exception that the electrons avoid being in the bond region together; the probability that they will appear there is even less than in the case where the two atoms are simply brought alongside with all interactions turned off.

1.3 Mixing of Functions: A Quantitative Description of Resonance

In the same manner that the hydrogen molecule can be described as a resonance mixture of covalent and ionic structures

$$H\text{---}H \leftrightarrow \overset{\oplus}{H} \ \overset{\ominus}{H} \leftrightarrow \overset{\ominus}{H} \ \overset{\oplus}{H} \tag{1-5}$$

the valence-bond wave function can be improved by adding an ionic term

$$\Psi^{\text{ionic}} = \frac{1}{\sqrt{2 + 2S^2}} \left[\phi_A(1)\phi_A(2) + \phi_B(1)\phi_B(2)\right] \times \frac{1}{\sqrt{2}} (\alpha_1\beta_2 - \beta_1\alpha_2) \tag{1-6}$$

to the previous covalent term (1-2). The mixing coefficient in

$$\Psi = \Psi^{\text{cov}} + \lambda\Psi^{\text{ionic}} \tag{1-7}$$

will be determined variationally, so as to optimize the total electronic energy. Although the two component wave functions (1-2) and (1-6) are far from orthogonal,[3] it is convenient to use the mixing coefficient to express the percentages of covalent and ionic character in the bond (in the present case 93 and 7% at equilibrium).

For more complicated systems, with a series of resonance structures I, II, III, and so on, the wave function will be chosen as a linear combination of the corresponding functions Ψ_I, Ψ_{II}, Ψ_{III}, and so forth:

$$\Psi = \sum a_n \Psi_n \qquad (1\text{-}8)$$

where Ψ_n is the wave function for a "canonical" valence-bond structure.* The function for a single structure is often quite complicated. It is an antisymmetrized product of atomic orbitals and spin orbitals,[4] which can be reduced to a combination of Slater determinants. Some examples are given in Table 1-1 for the simplest groups of atoms: two, three, or four. The resonance structures are shown alongside, where the notation $|\phi_A \bar{\phi}_B \phi_C|$ is the abbreviation for the determinant

$$|\phi_A \bar{\phi}_B \phi_C| \equiv \begin{vmatrix} \phi_A(1)\alpha(1) & \phi_B(1)\beta(1) & \phi_C(1)\alpha(1) \\ \phi_A(2)\alpha(2) & \phi_B(2)\beta(2) & \phi_C(2)\alpha(2) \\ \phi_A(3)\alpha(3) & \phi_B(3)\beta(3) & \phi_C(3)\alpha(3) \end{vmatrix} \qquad (1\text{-}9)$$

In the resonance structures dotted lines indicate triplet "bonding" between parallel spins, whereas $\overset{\uparrow}{A}$, $\vec{A}$, and $\overset{\downarrow}{A}$ indicate an atom with $S_z = \frac{1}{2}$, $S_z = 0$, and $S_z = -\frac{1}{2}$ electron, respectively.

In the case of *ionic resonance,* where charge can move back and forth between two centers, there are clearly two possible combinations of the corresponding functions:

$$\overset{\ominus\oplus}{AB} \leftrightarrow \overset{\oplus\ominus}{AB} \begin{cases} \phi_A(1)\phi_A(2) \pm \phi_B(1)\phi_B(2) & \text{(spatial)} \quad \text{or} \\ |\phi_A \bar{\phi}_A| \pm |\phi_B \bar{\phi}_B| \end{cases} \qquad (1\text{-}10)$$

For H_2, in equation (1-6), we took the positive combination, which alone had the correct spatial symmetry to mix with the covalent Heitler-London function. The energy of the combined function (1-7) is calculated by the variational technique; the secular equations yield two roots, the lower one of which is the improved ground state and the higher one, a high-lying $^1\Sigma_g^+$ ionic state whose first approximation is the positive combination (1-10). The negative combination (1-10) is the wave function for another ionic state of high energy, the lowest excited singlet state ($^1\Sigma_u^+$) of hydrogen. Thus it is generally useful to distinguish between "in-phase" resonance (positive combination) and "out-of-phase" resonance (negative combination) by appropriate signs written above the arrows:

$$\overset{\oplus\ominus}{AB} \overset{(+)}{\leftrightarrow} \overset{\ominus\oplus}{AB} \qquad \overset{\oplus\ominus}{AB} \overset{(-)}{\leftrightarrow} \overset{\ominus\oplus}{AB} \qquad (1\text{-}11)$$

---

*For a given system of atomic orbitals, the number of linearly independent wave functions is restricted to the so-called Rumer diagrams in which no lines intersect ("canonical" set).[4]

TABLE 1-1

Valence-Bond Wave Functions for Two, Three, and Four Atoms[a]

A B (two centers, two electrons)

One singlet	$\|\phi_A \bar{\phi}_B\| + \|\phi_B \bar{\phi}_A\|$	[identical with (1-2)]
One triplet	$\|\phi_A \bar{\phi}_B\| - \|\phi_B \bar{\phi}_A\|$	[identical with (1-4)]

A———B

$\vec{A} \cdots \cdots \overset{\rightarrow}{.B}$

A B C (three centers, three electrons)

Two doublets	$\|(\phi_A \bar{\phi}_B + \phi_B \bar{\phi}_A)\phi_C\|$
	$\|2\,\phi_A \phi_B \bar{\phi}_C + (\phi_A \bar{\phi}_B - \phi_B \bar{\phi}_A)\phi_C\|$

$\begin{array}{c} \overset{\uparrow}{C} \\ A \!-\!\!-\!\!-\! B \end{array}$

$\begin{array}{c} \overset{\downarrow}{C} \\ \overset{\uparrow}{A} \cdots\cdots \overset{\uparrow}{.B} \end{array}$

One quartet	$\|\phi_A \phi_B \phi_C\|$

$\begin{array}{c} \overset{\uparrow}{.C}. \\ \overset{\uparrow}{A}: \cdots : \overset{\uparrow}{.B} \end{array}$

A B C D (four centers, four electrons)

Two singlets

$$|(\phi_A\bar{\phi}_B + \phi_B\bar{\phi}_A)(\phi_C\bar{\phi}_D + \phi_D\bar{\phi}_C)|$$

$$|\phi_A\bar{\phi}_B\phi_C\bar{\phi}_D + {}_B\bar{\phi}_C\bar{\phi}_D\phi_A\bar{\phi}_B - \frac{1}{2}(\phi_A\bar{\phi}_B - \phi_B\bar{\phi}_A)(\phi_C\bar{\phi}_D - \phi_D\bar{\phi}_C)|$$

A———B
C———D

$\overset{\uparrow}{A}\cdots\overset{\uparrow}{B}$
$\overset{\downarrow}{C}\cdots\overset{\downarrow}{D}$

Three triplets (we write the $S_z = 1$ components)

$$|\phi_A\phi_B(\phi_C\bar{\phi}_D - \phi_D\bar{\phi}_C) - \phi_C\phi_D(\phi_A\bar{\phi}_B - \phi_B\bar{\phi}_A)|$$

$\overset{\uparrow}{A}\cdots\overset{\uparrow}{B}$ $\xleftrightarrow{(-)}$ $\overset{\uparrow}{A}\cdots\overset{\uparrow}{B}$
$\overset{\uparrow}{C}\cdots\overset{\uparrow}{D}$ $\overset{\uparrow}{C}\cdots\overset{\uparrow}{D}$

$$|\phi_A\phi_B(\phi_C\bar{\phi}_D + \phi_D\bar{\phi}_C)|$$

$\overset{\uparrow}{A}\cdots\overset{\uparrow}{B}$
C———D

$$|\phi_C\phi_D(\phi_A\bar{\phi}_B + \phi_B\bar{\phi}_A)|$$

A———B
$\overset{\uparrow}{C}\cdots\overset{\uparrow}{D}$

One quintet

$$|\phi_A\phi_B\phi_C\phi_D|$$

$\uparrow A\cdots\cdots B\uparrow$
$\uparrow C\cdots\cdots D\uparrow$

<hr>

"Dotted lines indicate "triplet coupling" and full lines, "singlet coupling" between electron pairs.

9

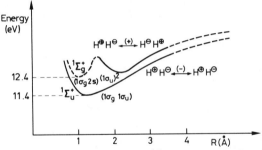

FIGURE 1-3.   The two lowest singlet "ionic" states in H_2.[6,*] (In practice, the dissociation limit is not H^+, H^- because of the lower energy of H· + H_{2s}^*). Zero of energy is ground-state H_2 at equilibrium geometry.

Out-of-phase resonance has also been called *antiresonance* (see Section 3.8).[5] The "in-phase" and "out-of-phase" ionic states of H_2 are shown in Fig. 1-3.

In a similar manner the resonance in benzene should be written

$$\begin{array}{ccc} \hexagon & \xrightarrow{(+)} & \hexagon \end{array} \tag{1-12}$$

for the ground state ($^1A_{1g}$), which is a positive combination of wave functions corresponding to the two resonance structures. The negative combination describes the lowest excited singlet state ($^1B_{2u}$).

1.4 Generalized Valence-Bond Method

Goddard[7-9] introduced a crucial improvement in valence-bond theory that makes it tractable† for multielectron systems to maintain conceptual simplicity. Certain selected electron pairs—the more labile electrons, in conjugated systems, or those that form bonds to be broken or made in reactions—are described by an ordinary valence-bond function

$$\Psi_{GVB} = (\chi_A(1)\chi_B(2) + \chi_B(1)\chi_A(2))\,(\alpha_1\beta_2 - \beta_1\alpha_2) \tag{1-13}$$

$$= |\chi_A\bar{\chi}_B| + |\chi_B\bar{\chi}_A|$$

[see (1-2) and the first line of Table 1-1]. But in (1-13), contrary to (1-2), orbitals

---

*On dissociation both zwitterionic states shortcut to atom-pair states in which one atom is excited to the $2s$ state. See Ref. 6 for the double-minimum behavior of the $^1\Sigma_g^+$ state.

†For the conventional valence-bond procedure and its complications in multielectron systems, see Ref. 10. For simplified valence-bond methods, see Ref. 11.

χ_A and χ_B are not just atomic or hybrid orbitals; they are solved for in a self-consistent manner. In this way the generalized valence-bond (GVB) orbitals are found to be essentially localized on one atom but partially delocalized on the other (Fig. 1-4). Wave function (1-13) is then, not surprisingly, a combination of traditional covalent (1-2) and ionic (1-6) wave functions. In fact, for hydrogen, the GVB energy is identical with that given by the best "covalent-plus-ionic" function (1-7). Hence Goddard introduces ionic character while preserving the simple "electron-pair" nature of the wave function.

In large systems the inner-shells and less important electrons are paired off in atomic or molecular orbitals (see Section 1.5). For the other electrons, all allowed pairing possibilities[4] should be considered. In practice, one writes out just one antisymmetrized product of functions similar to (1-13) for those centers that are paired in the most usual formula. This is the "perfect-pairing" GVB approximation. For example, to treat the π electrons of butadiene

$$
\begin{array}{c}
D \\
\diagup\!\!\!\diagup \\
\text{B}\!-\!\text{C} \\
\diagup\!\!\!\diagup \\
\text{A}
\end{array}
\qquad\qquad (1\text{-}14)
$$

only the first of the two singlets functions written out in Table 1-1 need be considered:

$$\Psi_{GVB} = |(\chi_A\bar{\chi}_B + \chi_B\bar{\chi}_A)(\chi_C\bar{\chi}_D + \chi_D\bar{\chi}_C)| \qquad (1\text{-}15)$$

It is that corresponding to "perfect-pairing" of atoms A and B and of atoms C and D. The second singlet function can be brought in at a later stage, as an additional variational term, with little improvement in the energy.

If the σ electrons of butadiene are included, Goddard may choose to ascribe to each σ bond a wave function that "correlates" the two σ electrons, as in (1-2) and (1-13). The total wave function then resembles (1-15), but with an antisymmetrized

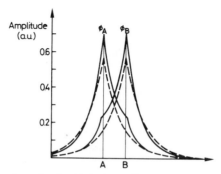

FIGURE 1-4. Generalized valence-bond orbitals for H_2.[9] Dashed lines indicate free atomic orbital. Reprinted with permission from W. A. Goddard, *J. Am. Chem. Soc.* **94**, 793 (1972). Copyright 1972 by American Chemical Society.

product of 11 successive terms (for the 9 σ pairs and 2 π pairs) such as $\chi_A\bar\chi_B + \chi_B\bar\chi_A$. Or he may elect to correlate just the π electrons and keep the σ electrons paired in molecular orbitals. The wave function is then

$$\Psi_{GVB} = |\sigma_1\bar\sigma_1\sigma_2\bar\sigma_2 \cdots \sigma_9\bar\sigma_9 \, (\chi_A\bar\chi_B + \chi_B\bar\chi_A)(\chi_C\bar\chi_D + \chi_D\bar\chi_C)| \quad (1\text{-}15a)$$

There is no guarantee, when such a trial wave function is used, that in the final solution the 18 paired electrons will all be σ electrons, the four "correlated" electrons π electrons. But if the orbitals are sufficiently flexible, this will be the case. Indeed, the energy gained by correlating π electrons (0.8 eV) exceeds that for CH σ electrons (0.4 eV) or for CC σ electrons (0.25 eV).

In cases where two or more bonding schemes are of comparable importance, as in allyl radical or benzene, the wave function that accounts for all possible (and simultaneously allowable)[4] spin pairings must be chosen.

1.5   Molecular Orbital Method

Instead of localizing the electrons in specific bonds, molecular orbital theory allows them to wander over the entire molecule. The delocalized orbitals are occupied by pairs and have characteristic energies that can be compared to a good approximation with experimental molecular ionization potentials. For the hydrogen molecule, the molecular orbital wave function is

$$\Psi_{MO} = \frac{1}{2 + 2S} [(\phi_A + \phi_B)\,(1)\,(\phi_A + \phi_B)\,(2)] \times \frac{1}{\sqrt 2} (\alpha_1\beta_2 - \beta_1\alpha_2) \quad (1\text{-}16)$$

where S has been defined in (1-3). Usually molecular orbital wave functions are given in determinantal form, such as

$$\Psi_{MO} = |\sigma_g\bar\sigma_g|, \qquad \sigma_g = \frac{1}{\sqrt{2 + 2S}}(\phi_A + \phi_B) \quad (1\text{-}17)$$

for H_2 or

$$\Psi_{MO} = |\psi_1\bar\psi_1\psi_2\bar\psi_2 \cdots \psi_n\bar\psi_n| \quad (1\text{-}18)$$

for a more complicated closed-shell system. In fact, the *orbital configuration* $\psi_1^2\psi_2^2 \cdots \psi_n^2$ is sufficient to define the wave function and to discuss the properties of the state.

The optimal shapes and the energies ϵ_i of the one-electron orbitals are given by the famous Hartree-Fock "self-consistent-field" (SCF) equations

$$\mathscr{F}\,\psi_i = \epsilon_i\psi_i, \qquad \mathscr{F} = h + \sum_{j=1}^{n}(2J_j - K_j) \quad (1\text{-}19)$$

where the Coulomb operator J_j and the exchange operator K_j are defined by

$$J_j(1)\psi_i(1) = \int \frac{\psi_j(2)\psi_j(2)}{r_{12}} d\tau_2 \, \psi_i(1)$$

$$K_j(1)\psi_i(1) = \int \frac{\psi_j(2)\psi_i(2)}{r_{12}} d\tau_2 \, \psi_j(1) \tag{1-20}$$

The one–electron operator $\mathscr{F}$ contains the effect h of the kinetic energy of the electrons considered together with the attractive potential energy of the nuclei, a smoothed-out Coulomb repulsion $\Sigma 2J_j$ between the electron in ψ_i and the other electrons, and a smoothed-out exchange term $-\Sigma K_j$ between the same electron and all other electrons with parallel spin. Roothaan[12] has shown how the Hartree-Fock equations can be solved in a self-consistent manner when, as usual, the molecular orbitals are expanded as linear combinations of atomic orbitals (LCAO approximation).

Since all the electrons are allowed to cover the entire molecular framework, the molecular orbital wave function has a large extent of ionic character when two or more electrons are simultaneously on the same atom. Such a defect can be remedied by changing the *restricted* nature of the wave function (1-17), with identical orbitals for electrons of opposite spin, into an *unrestricted* form with different orbitals for different spins.[13] For H_2, the correct unrestricted form is

$$\Psi = |\sigma_g \bar{\sigma}'_g| + |\sigma'_g \bar{\sigma}_g| \tag{1-21}$$

Functions of this type are commonly invoked for radicals since the different nature of the α and β spin orbitals makes allowance for the "exchange" or "spin" polarization due to the odd spin and in which inner-shell electrons of spin parallel to the odd electron are slightly attracted toward it (Section 7.3). For closed-shell molecules, a wave function such as (1-21), through modification of the correlation of electrons, serves to give the proper ratio of ionic and covalent character by decreasing the former. Indeed, the optimized form of (1-21) for H_2 is found to be identical with the best "covalent-plus-ionic" function (1-7) and with the optimized form of (1-13), the GVB wave function. It also dissociates to the correct limit $H\cdot$ + $H\cdot$, whereas the restricted determinant (1-17) does not.

In practical calculations the SCF equations for two-determinant functions such as (1-21) are quite complex (see Section 1-6). Therefore, conventional unrestricted wave functions are limited to a single determinant:

$$\Psi_{\text{unrestricted}} = |\psi_1 \bar{\psi}'_1 \psi_2 \bar{\psi}'_2 \cdots \psi_n \bar{\psi}'_n| \tag{1-22}$$

Unfortunately, such functions, although they do lead to tractable Hartree-Fock equations, are not any more proper eigenfunctions of the total spin. They contain small admixtures of functions appropriate to spin multiplicities different from that

TABLE 1-2

Energies of Restricted Single-Determinant Wave Functions

Wave Function	State	Energy[a]
$\|\psi_1\bar{\psi}_1 \cdots \psi_n\bar{\psi}_n\|$	Singlet	$2\sum_{j=1}^{n} h_j + 2\sum_{i=1}^{n}\sum_{j=1}^{n}(2J_{ij} - K_{ij})$
$\|\psi_1\bar{\psi}_1 \cdots \psi_n\bar{\psi}_n\psi_{n+1}\|$	Doublet	Same as Singlet $+ h_{n+1} + \sum_{j=1}^{n}(2J_{j,n+1} - K_{j,n+1})$
$\|\psi_1\bar{\psi}_1 \cdots \psi_n\bar{\psi}_n\psi_{n+1}\psi_{n+2}\|$	Triplet	Same as Doublet $+ h_{n+2} + \sum_{j=1}^{n}(2J_{j,n+2} - K_{j,n+2})$ $+ J_{n+1,n+2} - K_{n+1,n+2}$
	[singlet][b]	
$\|\psi_1\bar{\psi}_1 \cdots \psi_n\bar{\psi}_n\psi_{n+1}\psi_{n+2}\Psi_{n+3}\|$	Quartet	Same as Triplet $+ h_{n+3} + \sum_{j=1}^{n}(2J_{j,n+3} - K_{j,n+3})$ $+ J_{n+1,n+3} + J_{n+2,n+3} - K_{n+1,n+3} - K_{n+2,n+3}$
	[two doublets][b]	

[a] In the energy expressions h_j is the expectation value of the one electron operator h for the kinetic energy plus nuclear attraction energy, whereas J_{ij} and K_{ij} are the usual two-electron Coulomb and exchange integrals:

$$J_{ij} = \int \psi_i^*(1)\psi_i(1)\frac{1}{r_{12}}\psi_j^*(2)\psi_j(2)d\tau_1 d\tau_2$$

$$K_{ij} = \int \psi_i^*(1)\psi_j(1)\frac{1}{r_{12}}\psi_j^*(2)\psi_i(2)d\tau_1 d\tau_2$$

[b] There is an open-shell singlet state

$$|\psi_1\bar{\psi}_1 \cdots \psi_n\bar{\psi}_n\psi_{n+1}\bar{\psi}_{n+2}| + |\psi_1\bar{\psi}_1 \cdots \psi_n\bar{\psi}_n\psi_{n+2}\bar{\psi}_{n+1}|$$

that has the same outer *electronic configuration* $\psi_{n+1}^1\psi_{n+2}^1$ as this triplet. In certain calculations it may be convenient to estimate the energy of this two-determinant singlet from that calculated for the single-determinant triplet by using the molecular orbitals of the latter. This is possible by adding a simple correction term $+ 2K_{n+1,n+2}$ to the triplet energy. Similarly, one can obtain approximate energies of the two doublets that correspond to the same open-shell configuration as the quartet ($\psi_{n+1}^1\psi_{n+2}^1\psi_{n+3}^1$) by adding correction terms to the energy of the latter.

of the state under consideration. For instance, (1-22) written out for a singlet state contains some triplet-state character. Hence restricted functions are often preferred and are even used for energetic calculations of open-shell radicals.

Table 1-2 gives the energies for some of the simplest *restricted* single-determinant molecular orbital wave functions of the general type

$$\Psi_{\text{restricted}} = |\psi_1\overline{\psi}_1\psi_2\overline{\psi}_2 \cdots \psi_n\overline{\psi}_n\psi_{n\ +\ 1}\psi_{n\ +\ 2} \cdots| \qquad (1\text{-}23)$$

Even for such wave functions, the correct Hartree-Fock equations[14] are quite complex, and slight simplifications are sometimes introduced (see Section 1.11).

1.6 Configuration Interaction and Multiconfiguration SCF Method

In Section 1.5 we saw that relaxation of the constraint that electrons of opposite spin be paired in the same molecular orbital decreased the ionic character of the single-determinant molecular orbital wave function and lowered its energy. Hence for H_2, the function $|\sigma_g\overline{\sigma}'_g| + |\sigma'_g\overline{\sigma}_g|$ was an improvement over $|\sigma_g\overline{\sigma}_g|$. Another way to improve a molecular orbital wave function is to take note that a single "configuration" such as σ_g^2 can certainly not give an exact description of a state (even though it is the major component of that state and can be used to label the state). The correct description must involve the instantaneous correlation of electron motions, not the average correlation provided by the Hartree-Fock equations. A simple manner of doing this, even though it is not physically very illuminating, is to include in the wave function small admixtures of *configurations that correspond to excited states,* that is, that are the major components of such states and that serve to label them. These are called *excited* configurations, and the method is called *configuration interaction* (CI).[15] For a recent review of the intricacies of CI, see Ref. 16.

Hence in the present case of hydrogen, a convenient wave function improved over (1-17) will be

$$\Psi_{\text{CI}} = |\sigma_g\overline{\sigma}_g| + \lambda|\sigma_u^*\overline{\sigma}_u^*|, \qquad \sigma_u^* = \frac{1}{\sqrt{2 - 2S}}(\phi_A - \phi_B) \qquad (1\text{-}24)$$

where σ_u^* is the antibonding, empty molecular orbital of H_2 and $|\sigma_u^*\overline{\sigma}_u^*|$ is the wave function for the configuration σ_u^{*2} corresponding to the doubly excited singlet state of the molecule. Admixture of the singly excited $\sigma_g\sigma_u^*$ configuration is ruled out by symmetry. A CI calculation can then easily be "added on" to a Hartree-Fock SCF calculation since the Hartree-Fock equations (1-19) yield the necessary set of excited orbitals, albeit not very good ones.

One can even improve on this conventional CI calculation by looking for the best wave function of form (1-24) in which both orbitals σ_g and σ_u^* are optimized simultaneously. Hence the total energy is optimized in a self-consistent manner by varying λ, σ_g and σ_u^*—instead of varying only λ as in the conventional method.

The multiconfiguration SCF (MCSCF) equations thus obtained are far more complicated than the ordinary SCF equations.[17] When all the dust is settled, the best MCSCF function (1-24) is identical with the best unrestricted function (1-21) (as can be seen by writing $\sigma'_g = \sigma_g + \sqrt{\lambda}\sigma^*_u$), with the optimized GVB function (1-13), and with the best "covalent-plus-ionic" function of type (1-7). For larger systems, theoreticians will often restrict themselves to a CI calculation by using fixed orbitals; even with this restraint, the improvement in energy can be quite impressive, particularly when there are low-lying configurations available. Hence it is not surprising that much effort has been put recently into fast CI techniques.[18]

1.7   Basis Sets in SCF Calculations

Calculations that attempt to solve the exact Hartree-Fock SCF equations by using approximate molecular orbitals and improving on these *ad libitum* to obtain increasingly better SCF energies, are called *ab initio* calculations. Defects in calculated energies and properties can be traced back only to the quality of the wave function. Other methods, in which simplified forms of the SCF equations are used, are called *semiempirical* (see Sections 1.8 and 1.9).

In ab initio calculations the molecular orbitals are developed as LCAO's. The choice of "basis" atomic orbitals on each atom is thus crucial to the quality of a calculation. Generally the larger a set and the more "flexible," the lower the energy. The flexibility involves in particular the coexistence of rather contracted orbitals that nicely describe the electron near the nucleus and also somewhat diffuse orbitals allowing the electrons to behave appropriately far away from the nuclei (such diffuse orbitals are particularly important for describing anions).

The oldest type of basis set are the Slater orbitals,[19] which differ from exact hydrogenoid atomic orbitals by the lack of radial nodal surfaces:

$$1s = \sqrt{\frac{\zeta_1^3}{\pi}}\, e^{-\zeta_1 r}$$

$$2s = \sqrt{\frac{\zeta_2^5}{3\pi}}\, re^{-\zeta_2 r} \qquad (1\text{-}25)$$

$$2p = \sqrt{\frac{\zeta_3^5}{\pi}}\, r\cos\theta e^{-\zeta_3 r}$$

and so forth. The values of the Slater exponents ζ can either be chosen from Slater's own rules[19] or be required to optimize the energy of the system under consideration.

Major progress was made when Boys[20] suggested the use of Gaussian atomic orbitals. These simplify greatly the calculation of two-electron integrals involving r_{12} in the denominator. The basic Gaussian functions have the form

$$1s' = \left(\frac{2\alpha_1}{\pi}\right)^{3/4} e^{-\alpha_1 r^2}$$

$$2s' = \left(\frac{2\alpha_2}{\pi}\right)^{3/4} e^{-\alpha_2 r^2} \quad \text{(same spatial form as } 1s')$$

$$\vdots \qquad \vdots \tag{1-26}$$

$$2p'_z = \left(\frac{128\alpha_3^5}{\pi^3}\right)^{1/4} r\cos\theta e^{-\alpha_3 r^2}$$

$$3p'_z = \left(\frac{128\alpha_4^5}{\pi^3}\right)^{1/4} r\cos\theta e^{-\alpha_4 r^2} \quad \text{(same spatial form as } 2p_z)$$

$$\vdots \qquad \vdots$$

These Gaussian functions can first be used as a group of basis orbitals called *primitive* functions, provided that there are enough of them. They can also be used as fixed linear combinations, called *contracted* Gaussians.

Some of the earliest and most extensive sets of primitive Gaussian orbitals were provided by Huzinaga.[21],* A $(9s5p)$ set means that a set of nine Gaussian s orbitals with respective exponents $\alpha_1, \alpha_2 \ldots, \alpha_9$ and five Gaussian p orbitals with exponents $\alpha_{10} \cdots \alpha_{14}$ are sufficient to give good energies for the system under consideration. Since all five orbitals and all p orbitals have the same form [see (1-26)], there is no need to specify whether we are dealing with $1s$, $2s$, $3s \cdots$ or $2p$, $3p$, $4p \cdots$ orbitals, even for atoms of the second or third row.

By least-square fitting combinations of primitive Gaussian orbitals to Slater orbitals, Pople and his collaborators have built several very popular schemes of contracted orbitals. The STO-3G basis set uses a fixed linear combination of three Gaussians to fit conventional Slater orbitals.[23] For hydrogen and carbon, these combinations are, respectively,

Hydrogen: $\quad 1s_{(\zeta = 1.24)} = 0.1543s'_{(3.425)} + 0.5353s'_{(0.6239)} + 0.4446s'_{(0.1688)}$

Carbon: $\quad 1s_{(\zeta = 5.67)} = 0.1543s'_{(71.617)} + 0.5353s'_{(13.045)} + 0.4446s'_{(3.530)}$

$\qquad 2s_{(\zeta = 1.72)} = -0.09967s'_{(2.941)} + 0.3995s'_{(0.6835)} + 0.7001s'_{(0.2223)}$

$\qquad 2p_{(\zeta = 1.72)} = 0.1559p'_{(2.941)} + 0.6077p'_{(0.6835)} + 0.3920p'_{(0.2223)}$

$$\tag{1-27}$$

where the α values are written as subscripts below the Gaussian orbitals. Such a basis set—regardless of the number of Gaussians involved—is called a *minimal*

*A full list of references for atomic Gaussian basic sets can be found in Ref. 22.

basis since there are as many overall Slater orbitals as there are atomic orbitals to be populated in the molecule. An improvement would be to choose two Slater orbitals per atomic orbital, that is, a double-ζ basis set. However, even if each Slater orbital is expanded into only three Gaussians, there are six Gaussians per atomic orbital, yielding quite a large number of Gaussians overall. A *split-valence* basis set designated 4-31G was thus introduced[24] by using a minimal contracted 4G basis for the inner shells and double-ζ quality for the valence-shell orbitals.

A 3G contracted set represents the "inner part" and a single Gaussian, the "outer part" of such orbitals. For hydrogen 1s orbitals, there will be a similar 3G,1G couple. The energetic improvements of the 4-31G set over the STO-3G set are quite remarkable because of the greater flexibility of the valence-shell orbitals that can reach out to the neighboring atoms and account properly for molecular formation. Hence in methane CH_4

$$E(\text{STO-3G})^{23} = -39.7153 \text{ a.u.}^*$$

$$E(\text{4-31G})^{24} = -40.1395 \text{ a.u.} \qquad (1\text{-}28)$$

$$E(\text{Hartree-Fock})^{25} = -40.2142 \text{ a.u.}$$

Polarization functions (such as 3d orbitals for a first-row atom) may also be added to the basis set, but their role is not always fully understood.

1.8 Semiempirical Methods: PCILO, Extended Hückel, and X_α Methods

A great amount of energy has gone into simplifying the molecular orbital scheme, which becomes increasingly unwieldy for larger systems. Methods of this type are called *semiempirical,* and there is no guarantee that they provide upper bounds to the total molecular energy as previous methods do. However, the compensating ease of their calculational schemes has made them widely used in studying organic, inorganic, and biochemical systems. Relative to the conventional SCF molecular orbital method (Section 1.6), the simplification may occur at three levels: for the wave function, for the Hamiltonian operator, or for the integrals.

Simplification of the Starting Wave Function

There is a convenient way of generalizing the molecular orbital wave function (1-16) of H_2 to a larger molecule. Instead of writing a single determinant (1-17) where electrons are paired in *delocalized* molecular orbitals, one can keep the electrons paired one pair to a bond, each pair occupying a *localized* bond orbital. Such a wave function can in fact be obtained by transforming Ψ_{MO}, for instance, under the requirement that electron exchange between different orbitals be minimized.[26] The corresponding localized orbitals give the same electron density as do the correct delocalized molecular orbitals. The localized orbitals are concentrated

---

*Atomic units of energy.

on single bonds with slight tails on neighboring bonds. In the Perturbative Config-uration Interaction Using Localized Orbitals (PCILO) method[27,28] similar localized "bond orbitals," but with approximate expressions, are chosen at the outset and the electrons are placed by pairs in them. An antisymmetrized determinant of the bond orbitals is written out. The interaction between electrons in different bonds (in particular that due to the delocalization of an electron into a neighboring bond or the possible exchange of such electrons, which are a reflection of the tails obtained in the transformation of the exact function), is then included in a pertur-bation expansion. Slight amounts of "excited" bond orbitals are mixed in (to decreasing orders of magnitude of the interaction) with the ground orbitals as the energy is improved. Convergence of such expansions is, of course, much better in systems where bonds are not stretched too much. Although the PCILO method has been used mainly in conjunction with approximate methods of integral calculation (Section 1.9), it can also be used in an ab initio manner, keeping rigorously all molecular integrals.

Simplification of the Hamiltonian

An extremely simple method, introduced by Hückel for conjugated molecules[29] and applied by Wolfsberg and Helmholtz to inorganic complexes,[30] was extended by Hoffmann[31] to all hydrocarbons and later used for all organic and inorganic systems. An effective one-electron Hamiltonian is defined by its matrix elements between atomic orbitals ϕ_i. The diagonal matrix elements H_{ii} are obtained from ionization potentials or valence-state ionization potentials,* whereas the off-diag-onal matrix elements H_{ij} are given by the Wolfsberg-Helmholtz formula

$$H_{ij} = 0.5K(H_{ii} + H_{jj}) S_{ij} \qquad (1\text{-}29)$$

where S_{ij} is the atomic overlap integral. The value of K is arbitrarily chosen to be 1.75 or 2.00 eV. The extended Hückel method (EHT) has been surprisingly effec-tive in predicting relative conformational energies and favorable reaction paths, even if calculated geometries or absolute energies are poor. Its main weakness lies in the complete neglect of electronic interaction; in fact, two-electron terms are entirely absent from the theory that cannot provide, for instance, excitation ener-gies. But insofar as one-electron effects dominate much of chemistry, the method has proved to be a very powerful tool.

A more rigorous but still approximate Hamiltonian can be obtained if in the Hartree-Fock operator for a closed-shell

$$\mathscr{F} = h + \sum_{j=1}^{n} (2J_j - K_j) \qquad (1\text{-}19)$$

---

*One method is to perform a complete charge iterative calculation $H_{ii} = H_{ii} + \alpha\, q_i$ (q_i, charge on ϕ_i) until self-consistency is obtained.

the exchange part, which is the most difficult to calculate, is replaced by an average proportional to the total density ρ

$$-\sum_{j=1}^{n} K_j \rightarrow -\frac{9}{2}\alpha\left(\frac{3\rho}{8\pi}\right)^{1/3} \tag{1-30}$$

The term (1-30), when applied to and averaged over some molecular orbital ϕ_i, is easy to calculate. This is the so-called X_α method,[32,33] which is related to the early Thomas-Fermi model. Even with the simplification given in (1-30) the X_α one-electron equations are quite difficult to solve self-consistently for large molecules. To obtain simple functionals for the density, it is convenient to assume a spherical potential inside and a constant potential outside atomic spheres whose sizes become additional parameters ("muffin-tin" approximation). These numerous parameters constitute the main weakness of a method that, however, is particularly useful for molecules with a single heavy atom (XeF_6, $PtCl_4^{2-}$), where the spherical potential of the central atom is a dominant term.

In a similar line of thought, "pseudopotential" methods,[34] in which essentially the inner shells are replaced by model potentials, are being actively developed for systems with heavy atoms.

1.9   Semiempirical Methods: Neglect of Differential Overlap Methods[35]

A third route to simplification of the SCF equations is the approximate evaluation of certain molecular integrals. In this respect the most popular approximation has been the neglect of differential overlap in all integrals:

$$\phi_A^*(1)\phi_B(1) = \delta_{AB} \tag{1-31}$$

("zero-differential overlap") so that, for instance

$$\int\phi_A^*(1)\phi_B(1)\frac{1}{r_{12}}\phi_C^*(2)\phi_D(2)d\tau_1\,d\tau_2 = \delta_{AB}\delta_{CD}\gamma_{AC},$$

$$\gamma_{AC} = \int\phi_A^*(1)\phi_A(1)\frac{1}{r_{12}}\phi_C^*(2)\phi_C(2)d\tau_1\,d\tau_2 \tag{1-32}$$

and

$$\int\phi_A^*\phi_B\,d\tau = \delta_{AB} \tag{1-33}$$

(all overlap integrals vanish).

Introduced independently by Pople and by Pariser and Parr[36] in 1953 for the study of conjugated molecules, the approximation was used by Pople and Segal[37] (before being abandoned by its authors in favor of ab initio methods) to develop a simplified SCF method valid for all molecules, the complete neglect of differential overlap (CNDO) method.

In its most popular form, CNDO/2, this method[37] considers the two-electron integrals γ_{AC} as well as certain one-electron integrals (off-diagonal h_{ij}, diagonal h_{ii}) as parameters characteristic of the atoms involved. For instance, h_{AB} is chosen proportional to the true overlap between ϕ_A and ϕ_B, so that—as in the old Hückel theory—overlap actually does weigh in on the final results. Similarly, the one-center contribution from atom A involves the ionization potential and electron affinity of atom A. The major weakness of the theory lies probably in the neglect of explicit overlap integrals: although this failure is not too detrimental in the old Hückel theory, adapted for conjugated molecules in their equilibrium geometries, it becomes significant for bond-breaking or bond-making studies. On the other hand, the role of electrostatic interactions (Coulomb repulsion and attraction between net charges) seems to be overemphasized.

Nevertheless, major attempts have been made to improve on the analytical expressions for the parameters and their numerical values. The most prominent have been the "modified intermediate neglect of differential overlap" (MINDO) methods developed by Dewar and collaborators,[38] the most recent version of which is called MINDO/3.[39] One difference with the CNDO methods is the correct evaluation of one-center exchange integrals such as

$$\int \phi_A^*(1)\chi_A(1) \frac{1}{r_{12}} \phi_A^*(2)\chi_A(2) \, d\tau_1 d\tau_2 \neq 0 \tag{1-34}$$

which were made to vanish by (1-31). The conservation of one-center exchange was orginally introduced by Dixon.[40] Furthermore, the parameters are carefully chosen to give excellent agreement with experimental heats of formation, a valuable asset for accurate reproduction of enthalpies of reaction in the calculation of reaction paths. However, this method also falls under criticism because of the very large number of parameters and because its underlying mathematical structure does not seem to accommodate the subtle differences between similar bonds in different environments.[41]

The original "intermediate neglect of differential overlap" (INDO) method,[42] which preceded the MINDO method, has proved useful for calculations on free radicals since inclusion of (1-34) allows for "spin" or "exchange" polarization effects and the calculation of correct unpaired spin densities. Other, more recent, semiempirical methods[43] with approximations similar to those given previously here, have been oriented toward fitting better and better the ab initio Hartree-Fock matrix elements.

1.10 Energy Calculations that Use Spectroscopic Information

It is tempting to build the available spectroscopic information on the absolute energies, force constants, or even potential-energy shapes into a calculation of potential-energy surfaces. In 1929 London[44] proposed an empirical potential-energy function for a system of three hydrogen atoms, A, B, and C

$$E(r_1, r_2, r_3) = Q_{AB} + Q_{BC} + Q_{CA}$$
$$- (J_{AB}^2 + J_{BC}^2 + J_{CA}^2 - J_{AB}J_{CA} - J_{BC}J_{CA} - J_{AB}J_{BC})^{1/2} \tag{1-35}$$

which is the valence-bond result for the ground-state doublet of three hydrogen atoms if interatomic overlap is neglected. In (1-35) Q is the valence-bond Coulomb integral and J the *negative* valence-bond exchange integral.[45,*] Since $Q + J$ is the energy of $^1\Sigma_g^+$ ground H_2, whereas $Q - J$ is the energy of $^3\Sigma_u^+$ excited H_2 to the same approximation,[45] knowledge of the experimental potential curves for these two states of H_2 yields a potential for H_3, as shown by Eyring and Michael Polanyi[46] and Sato.[47] This so-called LEPS (London, Eyring, Polanyi, Sato) potential has been used in an extended form, introduced by John Polanyi and collaborators:[48,†]

$$E(r_1, r_2, r_3) = \frac{Q_{AB}}{1+a} + \frac{Q_{BC}}{1+b} + \frac{Q_{CA}}{1+c} - \left[\frac{J_{AB}^2}{(1+a)^2} + \frac{J_{BC}^2}{(1+b)^2} \right.$$
$$+ \frac{J_{CA}^2}{(1+c)^2} - \frac{J_{AB}J_{BC}}{(1+a)(1+b)} - \frac{J_{BC}J_{CA}}{(1+b)(1+c)} - \frac{J_{AB}J_{CA}}{(1+a)(1+c)} \right]^{1/2} \tag{1-36}$$

where a, b, and c are variables. These can either be chosen to vanish or be given values that generate series of similar surfaces, which are useful in dynamical studies.

The *diatomics-in-molecules* method[50,51] is based on a similar philosophy. It is a modified valence-bond theory in which the total energy is written as the sum of energies of all possible *diatomic fragments* (including, for each fragment, the energy of the separate atoms and their interaction energy), minus the sum of atom energies counted as many times as necessary to correct for their multiple inclusion in the diatomic fragment energy sum:

$$H_{total} = \sum_P \sum_{Q>P} H_{PQ} - (N-2) \sum_P H_P \tag{1-37}$$

Hence, for OH_2, the total energy can be expressed[50] in terms of the ground ($X^2\Pi$) and first-excited ($A^2\Pi$) state energies of OH and the $^1\Sigma_g^+$ and $^3\Sigma_u^+$ states of H_2, together with the energies of the oxygen atom (^3P, ^1D) and of the hydrogen atom. For H_3^+, important in studying the reaction $H^+ + H_2 \rightarrow H_2 + H^+$, we need[50,51] the ground-state energy of H_2 and the two lowest states (ground $^2\Sigma_g^+$, first excited $^2\Sigma_u^+$) of H_2^+. Of course, these diatomic potential curves must be known at all

---

*For the wave function, see Table 1-1.
†For a different procedure, see Ref. 49.

distances for the theory to be useful. The theory can be applied with or without interatomic overlap.

Force-field methods use experimental force constants, taken from infrared and Raman data, to estimate the energy changes due to increases or decreases in bond lengths and to angular deformations.[52–54] At the same time nonbonded repulsions are approximated by Lennard-Jones- or Buckingham-type potentials fitted to incontrovertible experimental data. For instance, the crucial hydrogen-hydrogen and carbon-carbon nonbonded interaction curves can be obtained[53] by fitting the experimental heat of sublimation of the hexane crystal and the experimental conformational energy of a methyl group of a cyclohexane ring (a 7.1–7.9-kJ/mole preference for the axial methyl over the equatorial methyl). The H—H potential is given as

$$E_{\underset{\text{nonbonded}}{\text{H—H}}} \text{ (kJ/mole)} = -0.460 \left(\frac{3.0}{r_{\text{Å}}}\right)^6 + 1.797 \exp\left(\frac{-r_{\text{Å}}}{0.221}\right) \tag{1-38}$$

and corresponds to a van der Waals radius of 1.5 Å. The difficulty with such methods, apart from the great number of numerical constants (really parameters) involved, is that they will fail for bond breaking or bond making in regions of the energy surface where the potential energy is not quadratic any more. Even in conformational calculations where the reliability is optimal, errors occur for the comparison of systems in which specific nonadditive effects (orbital interactions) exist. Of course, for systems with very large numbers of atoms, where other methods tend to become intractable, force field methods are the ultimate resource.

Finally, purely analytical potentials can be constructed[55] for stable triatomic molecules by fitting spectroscopic data for the molecule at equilibrium and, somewhat similarly to the method of diatomics–in–molecules, for various diatomic and monoatomic dissociation products.

1.11 Evaluation of Excited-State Energies *

In theory, excited-state energies should be calculated directly by applying the variational theorem to the appropriate state. Self-consistent field equations exist for such a problem and can be solved in principle for restricted wave functions of type (1-23).[14] The problem is tractable as long as orthogonality to the ground-state function is ensured through either the spin symmetry of the global excited function (e.g., ground singlet state, excited triplet state) or the spatial symmetry of the excited orbital (e.g., ground and excited singlet states of different spatial symmetry). However, if excited and ground states have the same spin-spatial symmetry, the solution becomes more complicated since orthogonality of the excited state to the ground state must be imposed in some manner. Otherwise, the computation

*For a pertinent review, see Ref. 56.

will yield simply the ground-state wave function, which has the lower energy of the two.

In practice, even when there is no orthogonality problem, the SCF equations[14] are quite difficult to solve because the electrons in the "open" part of the molecular shell feel a potential slightly different from those in the "closed" part. For instance, in a triplet

$$\Psi = |\psi_1\bar{\psi}_1 \cdots \psi_n\bar{\psi}_n\psi_{n+1}\psi_{n+2}| \qquad (1\text{-}39)$$

an orbital such as ψ_n feels on average a field $J - \frac{1}{2}K$ [see (1-20)] from either of the two open-shell orbitals ψ_{n+1} and ψ_{n+2} (since each of these has its spin parallel to *one* of the *two* electrons in ψ_n). On the other hand, an open-shell orbital such as ψ_{n+1} feels a field $J - K$ from the other open-shell orbital. Therefore, it is natural that Nesbet[57] should have proposed a single "compromise" Hartree-Fock Hamiltonian, in the present case one that reproduces quite closely the correct interaction for the majority of pairs

$$\mathscr{F} = H + \sum_{j=1}^{n} (2J_j - K_j) + (J_{n+1} - \frac{1}{2}K_{n+1}) + (J_{n+2} - \frac{1}{2}K_{n+2}) \quad (1\text{-}40)$$

(compare with (1-19)).

Often a series of excited-state energies are investigated, and it is more convenient (and nowadays more accurate than to approach these energies from a direct SCF calculation) to obtain these as the successive solutions of a configuration interaction calculation (Section 1.6). Each state is then a linear combination of excited configurations built on the *ground-state* orbitals, which is a significant imperfection since orbitals should rearrange appropriately for each state. Nevertheless, first approximations (and upper bounds) are obtained for the excited-state energies in this manner. As an example, Table 1-3 gives the diagonal and off-diagonal terms for one of the simplest configuration interaction matrices. This matrix involves the three singlet configurations built by exciting the electrons in the highest occupied molecular orbital of a closed shell. The off-diagonal matrix elements are modified if the orbitals are obtained in a different calculation, for instance, from a restricted open-shell calculation of the triplet (1-39) [by means of (1-40) and footnote *b* in Table 1-2]. Moreover, the actual *numbers* may be very different. They will certainly be better if one is aiming, for instance, at an accurate description of the open-shell ($^1\psi_n\psi_{n+1}$) singlet state.

In certain cases, if interest is focused on a *pair* of states, it may be appropriate to use a Hamiltonian that is intermediate between the proper Hartree-Fock or Nesbet-type Hartree-Fock Hamiltonians for the two states.[58] This is equivalent to creating a fictitious intermediate state with partial occupation numbers halfway between those of the two states under investigation.[59]

TABLE 1-3

3 × 3 Configuration Interaction Matrix for Three Singlets Obtained by Exciting from ψ_n into ψ_{n+1}.[a,b]

$^1(\ldots\psi_n^2)$	$^1(\ldots\psi_n\,\psi_{n+1})$	$^1(\ldots\psi_{n+1}^2)$
0	0	$K_{n,n+1}$
0	$h_{n+1} - h_n + \sum\limits_{j}^{n-1} [(2J_{j,n+1} - 2J_{j,n})$ $- (K_{j,n+1} - K_{j,n})] + J_{n,n+1} - J_{n,n}$ $+ K_{n,n+1}$	$\sqrt{2}(L_{n+1,n} - L_{n,n+1})$
$K_{n,n+1}$	$\sqrt{2}(L_{n+1,n} - L_{n,n+1})$	$2(h_{n+1} - h_n) + 2\sum\limits_{j}^{n-1} [(2J_{j,n+1}$ $- 2J_{j,n}) - (K_{j,n+1} - K_{j,n})]$ $+ J_{n+1,n+1} - J_{n,n}$

[a]All orbitals are solutions of Hamiltonian (1-19). Energy of the ground configuration is chosen as zero.

[b]In the matrix $L_{n+1,n}$ is the "hybrid" integral $\int \psi_{n+1}(1)\psi_{n+1}(1)\dfrac{1}{r_{12}}\psi_{n+1}(2)\psi_n(2)d\tau_1\,d\tau_2$. The integrals J and K have been defined in footnote a in Table 1-2.

References

1. M. Born and J. R. Oppenheimer, *Ann. Physik* **84,** 457 (1927).
2. W. Heitler and F. London, *Z. Physik* **44,** 455 (1927).
3. J. Braunstein and W. T. Simpson, *J. Chem. Phys.* **23,** 174, 176 (1955); H. Shull, *J. Appl. Phys.* (Suppl.) **33,** 290 (1962); I. Shavitt, personal communication to the author.
4. L. Pauling and E. B. Wilson, *Introduction to Quantum Mechanics* (McGraw-Hill, New York, 1935), pp. 374 sqq.
5. G. Levin and W. A. Goddard, *J. Am. Chem. Soc.* **97,** 1649 (1975).
6. R. S. Mulliken and W. C. Ermler, *Diatomic Molecules* (Academic, New York, 1977), Sections IIIB and VG; W. Kolos and L. Wolniewicz, *J. Chem. Phys.* **50,** 3228 (1969).
7. W. A. Goddard, *Phys. Rev.* **157,** 81 (1967).
8. F. W. Bobrowicz and W. A. Goddard in *Methods of Electronic Structure Theory,* edited by H. F. Schaefer, (Plenum, New York, 1977), Chapter 4.
9. W. A. Goddard, *J. Am. Chem. Soc.* **94,** 793 (1972).
10. M. Kotani, K. Ohno, and K. Kayama, *Handbuch der Physik,* edited by S. Flugge (Springer-Verlag, Berlin, 1961), XXXVII/2 p. 1 (Section 20).
11. C. Kubach and V. Sidis, *Phys. Rev.* **A 14,** 152 (1976); T. F. O'Malley, *J. Chem. Phys.* **51,** 322 (1969).
12. C. C. J. Roothaan, *Rev. Mod. Phys.* **23,** 69 (1957).
13. J. A. Pople and R. K. Nesbet, *J. Chem. Phys.* **22,** 571 (1954); J. C. Slater, *Phys. Rev.* **82,** 538 (1957).
14. C. C. J. Roothaan, *Rev. Mod. Phys.* **32,** 179 (1960).
15. E. A. Hylleraas, *Z. Physik* **48,** 469 (1928); H. M. James and A. S. Coolidge, *J. Chem. Phys.* **1,** 825 (1933).
16. I. Shavitt in *Methods of Electronic Structure Theory,* edited by H. F. Schaefer (Plenum, New York, 1977), Chapter 6.
17. A. C. Wahl and G. Das in *Methods of Electronic Structure Theory,* edited by H. F. Schaefer (Plenum, New York, 1977), Chapter 3.

18. B. O. Roos and P. E. M. Siegbahn, in *Methods of Electronic Structure Theory,* edited by H. F. Schaefer (Plenum, New York, 1977), Chapter 7; R. F. Hausman and C. F. Bender, ibid., Chapter 8; B. Brooks and H. F. Schaefer, *Internatl. J. Quant. Chem.,* **14,** 603 (1978); *J. Chem. Phys.* **70,** 5092 (1979).

19. J. C. Slater, *Phys. Rev.* **36,** 57 (1938).

20. S. F. Boys, *Proc. Roy. Soc.* **A200,** 542 (1950).

21. S. Huzinaga, *J. Chem. Phys.* **42,** 1293 (1965).

22. T. H. Dunning and P. J. Hay in *Methods of Electronic Structure Theory,* edited by H. F. Schaefer (Plenum, New York, 1977), Chapter 1.

23. W. J. Hehre, R. F. Stewart, and J. A. Pople, *J. Chem. Phys.* **51,** 2657 (1969).

24. R. Ditchfield, W. J. Hehre, and J. A. Pople, *J. Chem. Phys.* **54,** 724 (1971).

25. W. Meyer, *J. Chem. Phys.* **58,** 1017 (1973).

26. C. Edmiston and K. Ruedenberg, *Rev. Mod. Phys.* **34,** 457 (1963).

27. S. Diner, J. P. Malrieu, P. Claverie, and F. Jordan, *Chem. Phys. Lett.* **2,** 319 (1968).

28. J. P. Malrieu in *Semi-empirical Methods of Electronic Structure Theory,* Part A, edited by G. A. Segal (Plenum, New York, 1977), Chapter 3.

29. E. Hückel, *Z. Physik* **70,** 204 (1931); ibid. **72,** 310 (1931); **76,** 628 (1932).

30. M. Wolfsberg and L. Helmholtz, *J. Chem. Phys.* **20,** 837 (1952).

31. R. Hoffmann, *J. Chem. Phys.* **39,** 1397 (1963); R. Hoffmann and W. N. Lipscomb, ibid. **36,** 2179, 3489 (1962); ibid. **37,** 2872 (1962).

32. J. C. Slater in *The World of Quantum Chemistry,* edited by R. Daudel and B. Pullman (D. Reidel, Dordrecht, 1974), p. 3.

33. K. H. Johnson, *J. Chem. Phys.* **45,** 3085 (1966); J. C. Slater and K. H. Johnson, *Phys. Rev.* **B5,** 844 (1972); J. W. D. Connolly in *Semi-empirical Methods of Electronic Structure,* Part A, edited by G. A. Segal (Plenum, New York, 1977), Chapter 4.

34. J. C. Phillips and L. Kleinman, *Phys. Rev.* **116,** 287 (1959).

35. J. N. Murrell and A. J. Harget, *Semi-Empirical Self-Consistent-field Molecular Orbital Theory of Molecules* (Wiley-Interscience, New York, 1972); G. Klopman and R. C. Evans in *Semi-Empirical Methods of Electronic Structure Calculation,* Part A, edited by G. A. Segal (Plenum, New York, 1977), Chapter 2.

36. J. A. Pople, *Transact. Faraday Soc.* **49,** 1375 (1953); R. Pariser and R. G. Parr, *J. Chem. Phys.* **21,** 466, 767 (1953).

37. J. A. Pople and G. A. Segal, *J. Chem. Phys.* **44,** 3289 (1966).

38. N. C. Baird, M. J. S. Dewar, and E. Haselbach, *J. Am. Chem. Soc.* **92,** 590 (1970).

39. R. C. Bingham, M. J. S. Dewar, and D. H. Lo, *J. Am. Chem. Soc.,* **97,** 1285 (1975).

40. R. N. Dixon, *Molec. Phys.* **12,** 83 (1967).

41. J. A. Pople, *J. Am. Chem. Soc.* **97,** 5306 (1975); W. J. Hehre, ibid. **97,** 5308 (1975); M. J. S. Dewar, ibid. **97,** 6591 (1975).

42. J. A. Pople, D. L. Beveridge, and P. A. Dobosh, *J. Chem. Phys.* **47,** 2026 (1967).

43. T. A. Halgren and W. N. Lipscomb, *J. Chem. Phys.* **58,** 1569 (1973).

44. F. London, *Z. Elektrochem.* **35,** 552 (1929).

45. C. A. Coulson, *Valence,* 2nd ed. (Oxford University Press, 1961), Chapter 7.

46. H. Eyring and M. Polanyi, *Z. Physik. Chem. (Leipzig) B* **12,** 279 (1931).

47. S. Sato, *J. Chem. Phys.* **23,** 592 (1955).

48. P. J. Kuntz, E. M. Nemeth, J. C. Polanyi, S. D. Rosner, and C. E. Young, *J. Chem. Phys.* **44,** 1168 (1966).

49. R. N. Porter and M. Karplus, *J. Chem. Phys.* **40,** 1105 (1964).

50. F. O. Ellison, *J. Am. Chem. Soc.* **85,** 3540, 3544 (1963).

51. R. K. Preston and J. C. Tully, *J. Chem. Phys.* **54,** 4297 (1971); J. C. Tully in *Semiempirical Methods of Electronic Structure Calculation,* Part A, edited by G. A. Segal (Plenum, New York, 1977), Chapter 6.

52. F. H. Westheimer in *Steric Effects in Organic Chemistry,* edited by M. S. Newman (Wiley, New York, 1956), p. 523; J. B. Hendrickson, *J. Am. Chem. Soc.* **83,** 4537 (1961); K. B. Wiberg, ibid. **87,** 1070 (1965).

53. N. L. Allinger, M. A. Miller, F. A. Van Catledge, and J. A. Hirsch, *J. Am. Chem. Soc.* **89,** 4345 (1967), Table II.
54. S. Lifson and A. Warshel, *J. Chem. Phys.* **49,** 5116 (1968).
55. K. S. Sorbie and J. N. Murrell, *Molec. Phys.* **29,** 1387 (1975); S. Carter, I. M. Mills, and J. N. Murrell, *Molec. Phys.* **39,** 455 (1980).
56. A. Devaquet, *Topics in Current Chemistry,* Vol. 54 (Springer-Verlag, Berlin), 1975, Section I, p. 1.
57. R. K. Nesbet, *Rev. Mod. Phys.* **35,** 552 (1963).
58. L. Salem, C. Leforestier, G. Segal, and R. Wetmore, *J. Am. Chem. Soc.* **97,** 479 (1975).
59. M. J. S. Dewar, J. A. Hashmall, and C. G. Venier, *J. Am. Chem. Soc.* **90,** 1953 (1968); J. C. Slater and K. H. Johnson, *Phys. Rev.* **B5,** 844 (1972).

2
Potential-Energy Surface

The potential-energy surface is the cornerstone of all theoretical studies of reaction mechanisms. The topographic features of potential-energy surfaces are intimately associated with the experimentally observed features of the reaction process. A facile pathway linking the reactant to a specific product on the surface is synonymous with saying that this particular reaction actually occurs. An insurmountable mountain along another pathway indicates that another reaction is not feasible. Two cols of nearly equal energy are the theoretical translation of competing reactions in the test tube, and a shallow minimum nearby may confirm the existence of a postulated intermediate. It is therefore essential to know at the outset those properties that are general and common to *all* potential-energy surfaces.

2.1 General Form of Potential-Energy Surface

For a system of N nuclei, the potential-energy surface has $3N - 5$ dimensions ($3N - 4$ if the molecular system is to be linear throughout the reaction). Indeed, to the dimensions for each nuclear coordinate must be added the dimension that measures the total energy. Hence if diatomic molecules give simple potential-energy curves in a plane (Fig. 2-1a)[1], triatomic molecules such as H_2O or CO_2 already introduce problems since it is impossible to plot energy as a simultaneous function of bond angle and of both bond lengths. Two-dimensional contour lines with the energy as parameter are often used to give physically significant views of such potential surfaces (Fig. 2-1b)[2]. Generally, we are content with plotting a two-dimensional energy/coordinate section of the surface (Fig. 2-1c)[3]. This coordinate is chosen appropriately as that—or one of those—that undergoes the most important variations during the reaction. Optimally it will be the actual "reaction coordinate" (see Section 2.4). The advantage of such a two-dimensional description is the possibility of having the potential surfaces for *several* states on the same diagram. Sometimes there are two coordinates that both undergo important variations (Fig. 2-1d)[4].

We now investigate in more detail the cases of one (Fig. 2-1c) or two (Fig. 2-1d) important coordinates. In the first case, where a single nuclear coordinate is important, the profile for a reaction in the ground state generally has the characteristic form shown in Fig. 2-2a: a minimum for reactant(s), a barrier or a col, and a second minimum for product(s). The lowest col is discussed further in Section 2.3. Sometimes, however, there may be a *secondary minimum* along the path, as shown in Fig. 2-2b. This secondary minimum generally corresponds to an observable species called *an intermediate* (life-time $\geqslant 10^{-12}$) if the well depth is sufficiently large and the reactants do not have too large an internal energy due to vibrational excitation.[5,*] The profile shown in Fig. 2-2b might correspond, for instance, to the ionic dissociation of a molecule in polar solution (the intermediate then represents the ''intimate'' or ''tight'' ion pair whereas the product is the ''solvent-separated'' ion pair)[7] or to the isomerization of a 1,1,2-trisubstituted ethane from its least stable gauche form to its most stable gauche form through the gauche form of intermediate stability.[8]

When two nuclear coordinates of similar importance are involved in a reaction, it may be useful to plot the path of the reaction as a function of both of these. The energy then becomes a parameter that is used to draw *energy contours* in the two-dimensional framework defined by these coordinates (Fig. 2-3a). A possible pathway from reactant to product follows the reactant valley and climbs over the col into the product valley. In this particular example the first nuclear coordinate undergoes its transformation ''ahead'' of the second nuclear coordinate. Mechanistically, it may be important to know, and it may be possible to determine experimentally, which coordinate is involved at the beginning of the reaction and whether they transform quasisimultaneously or whether there is a time lag between them. This

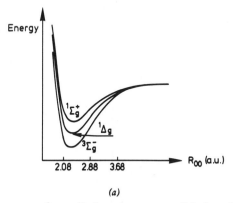

(a)

FIGURE 2-1. Potential-energy surfaces with the energy as an explicit dimension. *(a)* Dissociation of O_2 (potential curves for several states are shown).[1] Reprinted with permission from S. D. Peyerimhoff and R. J. Buenker, *Chem. Phys. Lett.* **16**, 235 (1972). Copyright 1972 by North-Holland Publishing Company.

---

*An intermediate, however, need not always correspond to a well-defined minimum (Fig. 2-2b). It may be very sloppy, corresponding to a large, relatively flat region of the potential surface—the ''twixtyl'' discussed by Hoffmann et al.[6]

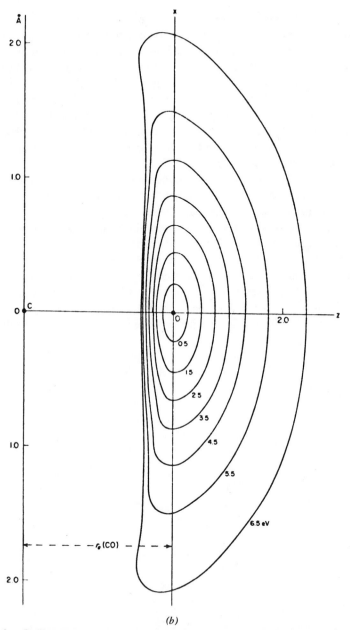

FIGURE 2-1. *(b)* Potential surface of the lowest singlet state of CO_2 as a function of the x and z coordinates of one O nucleus, keeping the rest of the molecule fixed on the abscissa *(z)* axis.[2] From *Electronic Spectra of Polyatomic Molecules* by G. Herzberg. Copyright 1966 by Van Nostrand. Reprinted by permission of the publisher.

30

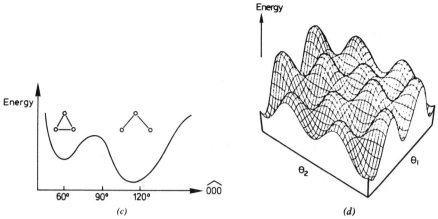

(c)   *(d)*

FIGURE 2-1. *(c)* Angular potential curve for ozone (the profile is a slice through the full four-dimensional surface).[3] Reprinted with permission from S. D. Peyerimhoff and R. J. Buenker, p. 213 of *The New World of Quantum Chemistry,* edited by B. Pullman and R. G. Parr. Copyright 1976 by D. Reidel Publishing Company. *(d)* Potential energy for topomerization of methyl formate (the two important coordinates are rotation angles).[4] Reprinted with permission from P. G. Mezey, p. 127 of *Applications of MO Theory in Organic Chemistry,* edited by I. G. Csizmadia. Copyright 1977 by Elsevier Scientific Publishing Company.

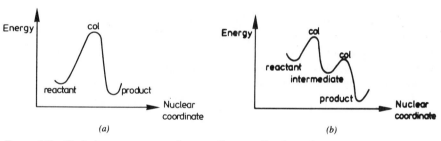

(a)   *(b)*

FIGURE 2-2. Typical energy-versus-nuclear coordinate profiles (ground-state systems): *(a)* no intermediate; *(b)* one intermediate.

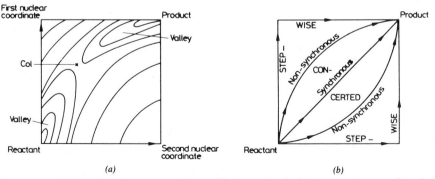

(a)   *(b)*

FIGURE 2-3. Case when two important nuclear coordinates are involved: *(a)* energy contours; *(b)* paths (general terminology).

31

question applies typically to cycloadditions

(2-1)

where two bonds are formed in a single reaction.

Figure 2-3*b* shows the terminology for the different possible paths:

A *"concerted"* path, where the two nuclear coordinates transform essentially together. The concert can occur in a perfectly *synchronous* manner or in a slightly *nonsynchronous* manner. Concerted pathways pass through the center of the diagram.

A two-step or *stepwise* path, where the two nuclear coordinates are transformed essentially one after another. Again, there may be some overlapping in time between the nuclear motions, or there may be a real time lag between the end of the first motion and the start of the second, depending on the extent to which the pathway follows the external boundary of the diagram (Fig. 2-3*b*).

In essence, a two-step reaction in time *generally* implies a valley in the energy contours at the top left or bottom right-hand corner of the contour diagram. Therefore, it will generally be associated with an *intermediate* of the type described in Fig. 2-2*b*. To a concerted process will correspond the profile shown in Fig. 2-2*a*, without any secondary minimum (sometimes called "energy concerted").[9, *]

The boundary between concertedness and stepwiseness is, of course, a hazy one.[†]

2.2   Arrhenius Rate Law: Activation Energy and Activation Entropy

The kinetic behavior of chemical reactions is almost universally governed by the Arrhenius rate law[11]:

$$\mathbf{k}\ (T)\ =\ A\ \exp^{-E/RT} \tag{2-2}$$

where A is the "preexponential factor" or "frequency factor," E is the measured or experimental activation energy for one mole of reacting substance, and R is Avogadro's number N multiplied by Boltzmann's constant. The few cases in which the Arrhenius equation is not obeyed are generally ascribed to tunneling.[‡]

*For cases where the degree of curvature of the pathway as a function of nuclear coordinates may not agree with indications from the potential-energy profile, see Ref. 10.

†There are, furthermore, stereochemical (preservation of stereochemical integrity) and kinetic (lower activation energy than that expected for a single rate-determining step) criteria of concertedness that may not agree with our theoretical description (J. A. Berson, private communication; see further in Chapter 4).

‡A recent instance is non-Arrhenius behavior in the bond shifting motion in pentalene, ascribed by J. Oth to carbon tunneling (J. Oth, private communication to the author); see also Ref. 12.

Über die Reaktionsgeschwindigkeit bei der Inversion von Rohrzucker durch Säuren.

Von

Svante Arrhenius.

Die Geschwindigkeit der Reaktionen, welche von Elektrolyten (Säuren und Basen) bewirkt werden, steht in einem sehr engen Zusammenhange mit der elektrischen Leitfähigkeit derselben. Aus theoretischen Gründen wurde Proportionalität zwischen diesen beiden Grössen angenommen[1]), und die nachher ausgeführten experimentellen Bestimmungen[2]) zeigten, dass dies annähernd der Fall ist. Jedoch konnte diese Beziehung nicht strenge aufrecht erhalten werden. Nachdem man die Möglichkeit gefunden hatte, den Dissociationsgrad zu berechnen, lag es nahe, zu versuchen, ob nicht Reaktionsgeschwindigkeit und Menge von wirksamen Jonen proportional seien, wie in erster Annäherung zu erwarten war; dies ist aber unterblieben, da die Beziehung auch in dieser Form keine streng gültige ist. Eine andere ins Auge springende Schwierigkeit, welche möglicherweise mit dem Mangel an Proportionalität zusammenhängt, ergiebt sich aus der Thatsache, dass die Reaktionsgeschwindigkeit (z. B. bei Inversion von Rohrzucker etc.), welche der Theorie nach von den Jonen des anwesenden Elektrolyten bewirkt wird, unter Umständen (durch Zusatz von Neutralsalzen) zunehmen kann, obgleich die Anzahl der wirksamen Jonen abnimmt. Da die Frage nach der Art des Zusammenhanges zwischen Anzahl der wirkenden Jonen und Reaktionsgeschwindigkeit von nicht unbedeutendem theoretischen Interesse ist, so habe ich versucht, diese Frage für einen der am meisten studierten Fälle, Inversion von Rohrzucker durch Säuren, unter Zuziehung von den vielen schon gemachten Versuchen zu einigen eigens für die Diskussion von mir neu ausgeführten einheitlich zu beantworten.

1. Einfluss der Temperatur auf die Reaktionsgeschwindigkeit.

Die Geschwindigkeiten der bisher in dieser Beziehung untersuchten Reaktionen werden sehr stark durch steigende Temperatur befördert.

[1]) Arrhenius: Bihang der Stockholmer Akad. **8**, Nr. 14, S. 60. 1884.
[2]) Ostwald: Journ. f. prakt. Ch. (2) **30**, 95. 1884; Lehrbuch **2**, 823. 1887.

Photostat of first page of the 1889 Arrhenius paper.

In a famous 1935 paper Eyring[13],* developed the theory of "absolute" reaction rates, in which an attempt was made to derive (2-2) directly from quantities available from the potential surface. Using statistical mechanics and defining the col on the surface as an activated "complex" or "state" ("the highest point along the lowest pass"[13]) to which he could apply thermodynamic conditions, Eyring obtained the rate constant **k** for various model reactions such as A + BC → AB + C or AB + CD → AC + BD. If E_0 is the height of the potential barrier, measured as the difference between the zero-point energy of the activated complex and that of the reactants, the relationship between the experimental activation energy E and E_0 (in molar units) is[15]

$$E = E_0 + \alpha RT \tag{2-3}$$

where α is a number (½ for a bimolecular collison) depending on the type of reaction. The additional ½ RT or so can be regarded as an extra translational energy, for the relative motion of the colliding partners, required to pass over the col. Indeed, Eyring treats the nuclear coordinate for reaction in the region of the col as a translational degree of freedom. In practice, "barrier height" and "activation energy" are often considered to be identical.

The experimental activation energy can also be related in an elegant manner[16] by collision theory to the average energy $\bar{E}^*$ of collisions in which reaction takes place

$$E = \bar{E}^* - \bar{E} \tag{2-3a}$$

where $\bar{E}$ is the mean energy of *all* collisions in the system. The averages are understood as statistical averages, with a Boltzmann factor. Again a slight variation with temperature of the experimental activation energy is uncovered, and the behavior of the two quantities $\bar{E}^*$ and $\bar{E}$ varies slightly with temperature.

A third, purely thermodynamic definition of activation energy exists by expressing the rate constant in terms of the *enthalpy of activation* $\Delta H^{\ddagger}$ and *entropy of activation* $\Delta S^{\ddagger}$ of the activated complex.[17] The result is[15,17]

$$\mathbf{k}(T) = \frac{kT}{h} e^{\Delta S^{\ddagger}/R} e^{-\Delta H^{\ddagger}/RT} \tag{2-4}$$

with (in molar units)

$$E = \Delta H^{\ddagger} + (1 - \Delta n^{\ddagger}) RT \tag{2-5}$$

where k is Boltzmann's constant and $\Delta n^{\ddagger}$ is the increase in the number of molecules

---

*For a very similar theory, including the assumption of equilibrium between reactant and transition state, see Ref. 14.

when the activated complex is formed from the reactants assumed to be perfect gases.

Equation (2-5) also shows that the preexponential or frequency factor A is related to the entropy of activation in a bimolecular reaction ($\Delta n^{\ddagger} = -1$)

$$A = e^2 \frac{kT}{h} e^{\Delta S^{\ddagger}/R} \qquad (2\text{-}6)$$

where $e = 2.718$. Hence the preexponential factor also depends on temperature. The fact that the preexponential factor and the exponential factor of reaction rates have strong variations in the entropy of activation $\Delta S^{\ddagger}$ and in the enthalpy $\Delta H^{\ddagger}$, respectively, is a warning that the theoretical interpretation of reaction rates cannot be built on the latter alone. If and when relative rates can be explained on the basis of relative potential barriers E_0 [see (2-3)], it must be assumed that variations in the entropy of activation in these reactions are small. As time unfolds, certain interpretations based on barrier differences may have to be revised and yield to interpretations based on order-disorder differences for the activated complex.

Note finally that activation volumes $\Delta V^{\ddagger}$, the changes in volume from reactants to activated complex, which require rate constant versus pressure measurements,[18] can give interesting mechanistic information[19] but have been the subject of little theoretical investigation.

2.3 Transition Structure or Transition "State": General Symmetry Properties

Instead of the term "activated complex" or "activated state" introduced by Eyring,[13] Evans and Polanyi[14] coined the term "transition state" for the lowest available col on the potential surface for the reacting species. The importance of this col had been noted as far back as 1910, both in Holland (the *"intermediate state"* described by Kohnstamm and Scheffer)[20] and in France (the "état critique" described by Marcelin).* The label "state" was justified[22] by the thermodynamic assumptions applied to the reacting system[17] in the region of the col. However, the col on the potential surface is really a specific geometric structure, that with the highest energy on the road from reactants to products. Thus the term "transition *structure*" seems more appropriate than "transition state" in spite of the historical tradition in favor of the latter. This choice eliminates confusion when the cols of different electronic states are considered ("ground-state transition state"; "excited-state transition state"). In this volume both terminologies are used freely.

Let us now be more precise about the properties of the transition state on the potential surface. The transition state is a local maximum or "saddle-point" on the surface. Hence, for a set of coordinates Q_i for which

$$\frac{\partial^2 E}{\partial Q_i \, \partial Q_j} = 0 \qquad \text{for} \quad i \neq j \qquad (2\text{-}7)$$

---

*Of the two Marcelin works cited here,[21] the latter is a posthumus paper published after Marcelin died on the war front.

where E is the energy (Section 1.1), the following conditions must be satisfied:

$$\frac{\partial E}{\partial Q_i} = 0 \qquad \text{for all coordinates } i$$

$$\frac{\partial^2 E}{\partial Q_r^2} < 0 \qquad \text{for a certain coordinate } Q_r \tag{2-8}$$

$$\frac{\partial^2 E}{\partial Q_j^2} > 0 \qquad \text{for all other coordinates } Q_j (j \neq r)$$

The coordinate Q_r along which the transition state is a maximum is the *reaction coordinate*. Along all the other coordinates the transition state is a minimum. A natural choice[23,24] of coordinates to study reaction paths near the transition state is then the ensemble of normal coordinates, between which there are no off-diagonal terms [see (2-7)] in the quadratic expansion of the potential energy. The normal coordinates also diagonalize the kinetic energy. Commonly used internal coordinates (bond lengths and bond angles) cannot be used to define the transition state because they seldom satisfy (2-7). It should be noted that the solutions Q_i to (2-7) and (2-8), like all normal coordinates, depend on the *masses* of the atoms.

Murrell and Laidler have shown[23] that a transition structure can be on only one reaction coordinate even when there are several symmetry-equivalent reactants or products. Indeed, if there were two orthogonal coordinates for which the structure is a local maximum, a different coordinate with lower activation energy could easily be found. An immediate consequence is that certain transition states of high symmetry (degenerate saddle points such as "monkey" saddle-points[4]) are ruled out. This is the case of triangular D_{3h} structures such as would occur in

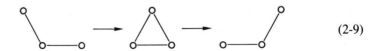

$$\tag{2-9}$$

Here the curvature of the surface is found to be positive along both e' normal coordinates of the triangle.[25] This triangle is thus a secondary minimum, and not the transition structure, which must be asymmetrical. Approximate energy contours are shown in Fig. 2-4. Similarly, in the $H + H_2$ exchange reaction, triangular H_3, although a hill, is not the transition structure. Consequences of this result on the appropriate statistical factors (number of equivalent forms of the activated complex that can arise from the reactants) have also been examined.[26,*] Hence care must be exercised when proposing transition structures of high symmetry.

Further considerations have been applied[28] to the transition "vector," the direction of most negative curvature (Q_r) at the transition state. First of all, as shown

---

*Also, Pollack and Pechukas[27] have recently shown that symmetry numbers, not statistical factors, should be used in the theory of absolute reaction rates.

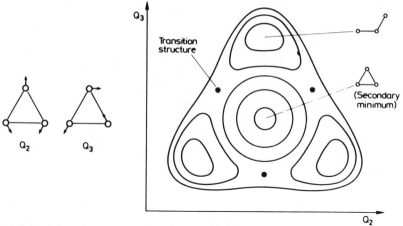

FIGURE 2-4.   Schematic energy contours for ozone.[25] The coordinates are the two E' coordinates. Reprinted with permission from J. N. Murrell, p. 161 of *Quantum Theory of Chemical Reactions*, Vol. 1, edited by R. Daudel, A. Pullman, L. Salem, and A. Veillard. Copyright 1980 by D. Reidel Publishing Company.

previously, it cannot be a degenerate mode. But also, if there is any symmetry operation that converts one form of product to another form of product, the transition vector must be unchanged under the operation. For instance, in the conformational change from chair cyclohexane to twist-boat cyclohexane

$$\text{(2-10)}$$

invariance of the chair under reflection in the plane of the diagram shown in (2-10) requires that the transition vector belong to the a_1 representation of the C_s point group. Also, if there is any symmetry operation that converts product into reactant, the transition vector must change sign under that operation. For instance, the substitution reaction $H + CH_4 \rightarrow CH_4 + H$ could be surmised to occur with inversion (actually a high-energy process):

$$\text{(2-11)}$$

However, inversion cannot occur through the C_{4v} transition state because there is no representation of the C_{4v} point group that is antisymmetric under the squared operation C_4^2 that relates reactant to product.

Finally, Pechukas[29] has shown that the symmetry of a nuclear configuration must be conserved along a path of "steepest descent" (see Section 2.4) so long as the path does not reach a stationary point. It follows that if a transition state is a saddle-point linked directly with reactants and products by paths of steepest descent, its symmetry is limited to the joint symmetry of reactant and product. Returning to (2-10), it would appear at first sight that this theorem rules out the C_s "sofa" structure as transition state in the passage from D_{3d} chair form to D_2 twist-boat form:

$$D_{3d} \qquad\qquad C_S \qquad\qquad C_{2v} \qquad\qquad D_2$$

$$(2\text{-}12)$$

However the C_s structure is not *directly* linked to the D_2 product by steepest descent—only to the C_{2v} boat structure, from which two pseudorotation steepest-descent paths of new symmetry fork out to the two twist-boat products. Hence the C_s sofa remains a potential transition state.

2.4   Various Types of Reaction Coordinate

The term reaction "path" or reaction "coordinate" covers a number of different possibilities.[4,30,31]

Steepest-Descent, or Least Steepest-Ascent, Pathway[32,33] (Fig. 2-5)

Starting at the transition structure, one moves toward the reactant valley or the product valley along the steepest slope. The path is constantly orthogonal to the energy contours, in the direction of the negative gradient of the potential, except at the transition structure itself, where the direction Q_r of most negative curvature is chosen. If a local perpendicular is drawn to this path, the potential energy along the perpendicular has a minimum at the point where the perpendicular intersects the steepest-descent path.

Figure 2-6 illustrates the steepest-descent pathway[31] in the extremely simple case where the contours are circles. The starting point x_0, y_0 has energy E_0 (2.24 units). Near E_0 the change in energy has the form

$$\Delta E = m_x \, \Delta x_0 + m_y \, \Delta y_0 \qquad\qquad (2\text{-}13)$$

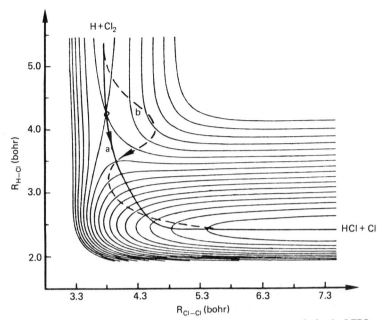

FIGURE 2-5.   Steepest-descent pathway *(a)* and a minimum energy pathway *(b)* for the LEPS potential-energy surface of H + Cl$_2$ → HCl + Cl.[33] Reprinted with permission from N. M. Witriol, J. D. Stettler, M. A. Ratner, J. R. Sabin, and S. B. Trickey, *J. Chem. Phys.* **66**, 1141 (1977). Copyright 1977 by American Institute of Physics.

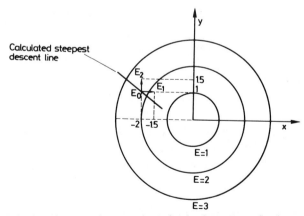

FIGURE 2-6.   Determination of steepest-descent pathway for circular contours. Starting point is E_0 from which small increments lead to E_1 and E_2.

where two separate explorations to points 1 ($E_1 = 1.80$) and 2 ($E_2 = 2.50$) have been required to obtain m_x (-0.88) and m_y (0.52). The steepest-descent line in the region of (x_0, y_0) is then defined by

$$\Delta y = + \frac{m_y}{m_x} \Delta x \qquad (2\text{-}14)$$

whose slope (-0.59) is close to the correct value (-0.50) that would obtain if infinitesimal increments had been used. The contour tangent has the equation

$$\Delta y = - \frac{m_x}{m_y} \Delta x \qquad (2\text{-}15)$$

The steepest-descent ''path'' would require the repetition of such a calculation at a neighboring point on the steepest-descent line, and so forth in such a way as to avoid accumulation of errors (see Section 2.5).

Intrinsic Reaction Coordinate[34]

This is also a steepest-descent pathway, which is required to pass through reactant, transition state, and product.

''Minimum Energy'' Pathways[32,33,35]

These are essentially pathways that choose a given reaction coordinate and minimize the energy relative to the other coordinates. For instance, if there are two important coordinates x and y, one satisfies the conditions

$$\frac{\partial E(x,y)}{\partial y} = 0, \qquad \frac{\partial^2 E(x,y)}{\partial y^2} > 0 \qquad \text{for a trial point} \quad x > x_c \qquad (2\text{-}16a)$$

where x_c is the coordinate of the assumed position of the col and

$$\frac{\partial E(x,y)}{\partial x} = 0, \qquad \frac{\partial^2 E(x,y)}{\partial x^2} > 0 \qquad \text{for a trial point} \quad x \leqslant x_c \qquad (2\text{-}16b)$$

Such reaction paths may lead to kinks or even discontinuities and in particular may differ wildly from the steepest-descent path (Fig. 2-5). The latter is truly *the* minimum-energy path or the least-energy pathway.

Classical Trajectories

The previous paths correspond to a set of atoms moving infinitely slowly (zero momentum) on the potential surface. Trajectories, on the other hand, are obtained by solving equations of motion for the reactants, with appropriate initial conditions including initial momenta for the nuclei. As a consequence, dynamic effects, such as the centrifugal force, will make the molecule go along pathways other than the

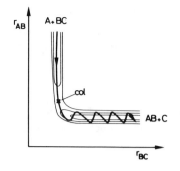

FIGURE 2-7. Dynamical trajectory on a potential-energy surface for an exothermic A + BC → AB + C reaction with an early barrier.[36] Reprinted with permission from J. C. Polanyi, *Acc. Chem. Res.* **5**, 161 (1972). Copyright 1972 by American Chemical Society.

steepest descent. The trajectories may be quite lengthy (see Fig. 2-7[36]) as the molecule rebounds on ridges and shoots down slopes, as would a billiard ball. The steepest-descent path in this respect becomes the *"minimum reaction path"*.[37]

2.5 Gradient and Other Methods for Calculating Reaction Paths

Finding the exact location of the transition state [because of the first of equations (2-8)] and drawing precisely the steepest-descent reaction path (because it follows the direction of negative gradient of E) both require accurate calculations of the energy gradients $\partial E/\partial Q_i$. Methods for *analytical* evolution of such gradients have been developed, particularly under the impulse of Pulay.[38–43] Of special interest is the use of gradient calculations for obtaining reaction coordinates.

Although we discuss here paths for systems with zero momentum, the influence of the atomic masses can be included indirectly by choosing mass-weighted Cartesian coordinates

$$x_i = \sqrt{M_i}\, X_i \tag{2-17}$$

or else can be entirely excluded by choosing the Cartesians

$$X_i, \ldots X_n \tag{2-18}$$

themselves. In the former case the steepest-descent pathway or intrinsic reaction coordinate (Section 2.4) will be slightly different, for instance, for the two systems $T + CH_4^-$ and $H + CH_4^-$.* Each pathway corresponds to a dynamical trajectory where the speed of the particle would be put back to zero after each infinitesimal advance.[43]

---

*The author is grateful to C. Leforestier for a discussion of gradient calculations. See in particular Figs. 6 and 7 of Ref. 43.

Regardless of the coordinate system used, the methods for obtaining the steepest-descent path or some approaching reaction coordinate fall into two categories: methods that start at the transition state and evolve to reactant and products; and methods that link reactant and product.

Methods that Start at the Transition State

In such methods the calculation has two parts: (1) determination of the transition state; and (2) determination of the pathway to reactants and products. For location of the transition state, the most common technique is to minimize the *gradient norm*[39,44,45]

$$|\mathbf{g} \cdot \mathbf{g}| = |\nabla E \cdot \nabla E| \qquad (2\text{-}19)$$

The difficulty is that such a minimization can lead to any stationary point, minimum or maximum, as illustrated in Fig. 2-8 in the two-dimensional case[45] (in certain cases it can even lead to nonstationary points[44]). Hence the search may be difficult if there are secondary minima in the neighborhood of the required saddle-point. Sometimes symmetry can be used directly to optimize the structure of an intermediate that should plausibly lie very close to the transition structure.[46]

Once the transition structure or an approximation thereof is found, the procedure for obtaining the steepest-descent pathway is straightforward[47] and starts off as in the simple case described in Section 2.4. A small vectorial move is made along the gradient of the potential

$$\mathbf{Q}_1 = \mathbf{Q}_0 - a \text{ grad } E_0 \qquad (2\text{-}20)$$

where a is an adjustable parameter and $\text{grad } E_0$ is evaluated at the starting point

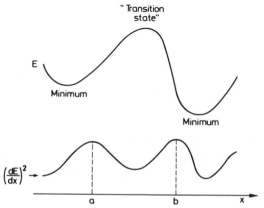

FIGURE 2-8.   A gradient norm search may lead to minima (if x has been chosen outside the interval a,b) instead of the required transition state.[45] Reprinted with permission from D. Poppinger, *Chem. Phys. Lett.* **35**, 550 (1975). Copyright 1975 by North-Holland Publishing Company.

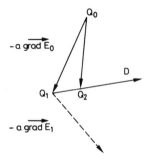

FIGURE 2-9. Gradient procedure for obtaining a small vectorial approximation to the reaction coordinate.[47]

$\mathbf{Q}_0$. However, to *compensate* for the possible change in this gradient between $\mathbf{Q}_0$ and the next point $\mathbf{Q}_1$, the vector $\mathbf{Q}_0\mathbf{Q}_1$ is made to *swing back* along the direction

$$\mathbf{D} = \frac{\mathbf{grad}\ E_1}{|\mathrm{grad}\ E_1|} - \frac{\mathbf{grad}\ E_0}{|\mathrm{grad}\ E_0|} \qquad (2\text{-}21)$$

where $\mathbf{grad}\ E_1$ is evaluated at point $\mathbf{Q}_1$. Three points are calculated along $\mathbf{D}$ until an energy minimum is found by parabolic fit at a point $\mathbf{Q}_2$. The vector $\mathbf{Q}_0\mathbf{Q}_2$ is considered to be a small fraction of the reaction coordinate (Fig. 2-9).

Methods that Link Reactant and Product

Such methods are based on the philosophy that the least-energy pathway must in some continuous manner link reactant and product. The simplest such method chooses arbitrarily one or two degrees of freedom as independent variables going from their reactant value to their product value and minimizes the energy with respect to the remaining coordinates along the path. This gives essentially a "minimum energy path" as defined in Section 2.4, with its inherent defects. For a strongly constrained system (molecule undergoing internal rotation), such a procedure is reasonable. But for looser, more complex reactants, artifacts may occur, such as abrupt changes in the nature of the optimized variable in the two-variable case* (see also Fig. 2-5, path *b* in the middle) or blind alley paths with apparent "hysteresis" effects[48] that are due essentially to the poor choice of variables.

Other methods attempt to use some form of interpolation, generally linear, between reactants and products. First, the internal coordinates shared by reactants and products may be varied linearly and simultaneously from reactant to product values,[49] leading to a *"linear internal coordinate"* path that smoothly connects reactant and product. A difficulty occurs for bimolecular reactions, in which certain

---

*Sudden changes that may be artifacts of this type include the sudden CH_2-group rotation in the conrotatory ring closure of butadiene to cyclobutene (see Halgren and Lipscomb[50] for the accurate reaction coordinate) and the sudden opening of the CH_3 umbrella in the $H^- + CH_4$ substitution reaction (see Ishida et al.[47] for the accurate reaction coordinate).

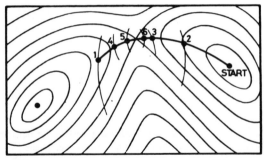

FIGURE 2-10.   Localization of a saddle-point by generating a series of ascending valley points.[30] Note that points 2 and 3 are both generated from 1 and 4 and 5 both from 3. Reprinted with permission from K. Müller, *Angew. Chem. Internatl. Ed.* **19,** 1, (1980). Copyright 1980 by Verlag Chemie GMBH.

coordinates of the product are infinite at the start but must be given finite values. Alternatively all the internuclear distances themselves may be varied linearly to give a *"linear synchronous transit"* pathway.[50] Some form of optimization in a direction perpendicular to these paths is then used to go from the highest point[50] on the trial path, or from that point where the energy gradient goes through a secondary minimum,[49] to the actual transition state that is assumed to lie nearby.

Other procedures have been proposed to obtain the least-energy pathway.[51,52] Müller and Brown[52] have developed a method of locating the transition state by starting at a valley point and drawing a hypersphere (sufficiently large so that it goes beyond the assumed position of the transition state) that intersects the hypersurface. The minimum along the intersection curve determines the new starting point (Fig. 2-10) from which one draws a second hypersphere, and so forth. The numerical effort yields not only the transition state, but also a set of points that are a rough guide to the reaction path. However, Müller's method itself fails if the transition state falls outside the search range (it is not in the assumed region "between" reactant and product) or when the constraint curve passes through the transition state at a bad angle, along a direction in which the energy is a maximum. To quote James McIver,* in his opinion "there does not yet exist a tractable, automatic and foolproof method of finding a transition state."

2.6   Principle of Least Motion†

In 1924 in France, Muller and Peytral, who had been studying the pyrolysis of small organic molecules,[53] enunciated[54] the "principe de la moindre déformation moléculaire" (principle of least molecular deformation), which stated that a reaction occurs in such a fashion that the atomic linkages differ as little as possible between initial and final state. In 1938 Rice and Teller[55] stated the "principle of

*Private communication.
†This section has been partly inspired by an unpublished review by G. Brodsky, a graduate student at Harvard University (1972).

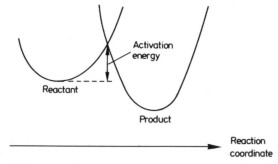

FIGURE 2-11. Potential-energy surfaces around reactant and product and their intersection. (The activation energy at the intersection can further be lowered by resonance, i.e., by the electronic "least-motion" effect.)

least motion," giving conditions that would favor a low activation energy: least motion of the atoms, with as little change as possible in their position; and least motion of the electrons, in which the electron orbits should not change too much to transform from the old to the new valence picture.

In a sense the second proposal covers, in a vague fashion, the entire problem of electronic behavior at transition states and evokes the fact that resonance at the transition state between two closely looking structures, one resembling reactant and one resembling product, should lower the activation energy. A good part of this book is devoted to studying the details of such electronic behavior, so we do not proceed further with this aspect.

Concerning the atoms, Rice and Teller illustrate the principle of least motion by considering the potential energies around reactant and product, along the reaction coordinate, as parabolic curves. The closer the two minima, that is, the shorter the pathway, the lower is their intersection (Fig. 2-11). The difficulty with such a construct is that on a multidimensional potential surface the straight-line pathway directly linking reactant to product is seldom the least-energy pathway. This is clear, for instance, in the unimolecular decomposition of cyclobutane, which does not break up by having the ethylene moieties simply move apart:

$$(2\text{-}22)$$

Even in a triatomic reaction

$$A + BC \rightarrow AB + C \qquad (2\text{-}23)$$

the principle would seem to want bond BC to start stretching as A approaches, thus circumventing the need for A to climb the full repulsive potential along its approach

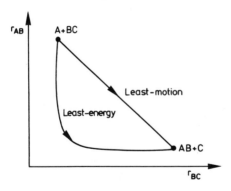

FIGURE 2-12.   Least-motion versus least-energy pathway in a triatomic substitution reaction. See Figs. 2-5 and 2-7 for typical energy contours.

to BC (high intersection). Instead of this assumed diagonal pathway from reactant to product (Fig. 2-12), known energy contours always favor the roundabout pathway. Similarly, famous examples[56] of reactions yielding the thermodynamically least favored product such as

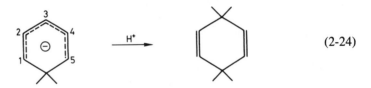

(2-24)

seem to be due* simply to the larger charge (and larger available orbital amplitude) on atom 3.

Recently efforts have been made to put the least-motion principle on a more mathematical basis.[56,58,59] Ehrenson[59] minimizes the sum of the squares of the mass-weighted Cartesian atomic distortions d_i:

$$D_m^2 = \sum_{i=1}^{N} m_i d_i^2 = \sum_{i=1}^{N} m_i \{(x_i' - x_i)^2 + (y_i' - y_i)^2 + (z_i' - z_i)^2\} \quad (2\text{-}25)$$

where the primed coordinates refer to product and the unprimed coordinates, to reactant. An extended, dynamical least-motion model has also been developed.[60]

Altogether, in spite of its obvious flaws, it is surprising that the least-motion principle still predicts correctly the conrotatory closure of butadiene to cyclobutene, or the anti elimination of olefins to acetylenes.[58] It has been suggested that the principle of least motion works when the behavior of the nuclear repulsion energy

*P. Hiberty and Y. Jean, unpublished observations. For a more refined observation based on molecular potentials (Section 6.10), see Ref. 57.

is parallel to that of the total energy: definite conditions for this to occur have yet to be established.*

2.7 Substituent and Steric Effects on Transition-State Geometry

The assumption that substituent effects can be considered as producing *linear* energy perturbations at the transition state has led to an important set of conclusions called *Thornton's rules*.[62] Thornton first distinguishes, in a reaction, the "parallel" coordinate—essentially the direction of the reaction coordinate—and the other, "perpendicular" coordinates at the transition state. For a reaction

$$A + BC \rightarrow AB + C \qquad (2\text{-}23)$$

the usual energy contour diagram (Fig. 2-13[62]; see also Figs. 2-5 and 2-7) shows the parallel coordinate to correspond roughly to the antisymmetric stretching coordinate of the triatomic system ABC:

$$A \rightarrow\; \leftarrow B - C \rightarrow \qquad (2\text{-}26)$$

The single perpendicular coordinate is the symmetric stretching coordinate

$$\leftarrow A - B - C \rightarrow \qquad (2\text{-}27)$$

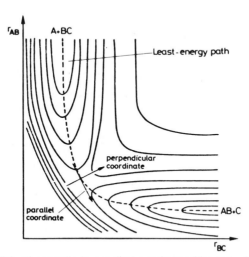

FIGURE 2-13. Parallel and perpendicular coordinates at the transition state of a triatomic substitution reaction.[62] Reprinted with permission from E. R. Thornton, *J. Am. Chem. Soc.* **89,** 2915 (1967). Copyright 1967 by American Chemical Society.

*Csizmadia et al.[61] have demonstrated an "antiphase" rule in which nuclear and electronic energies [corresponding respectively to V_{nn} and the other terms in (1-1)] have opposite phase along conformational coordinates. Here the purely nuclear term must dominate the terms that are partially or totally electronic.

For polyatomic systems, the same notions will apply to the normal coordinates Q_i at the transition state (Section 2.3).

Let us now substitute an atom or group in the reacting system. To the *parabolic* potential energy for the parallel ($Q_\parallel$) and perpendicular ($Q_\perp$) normal coordinates at the transition state we algebraically add a *linear* energy. Figure 2-14 shows the result when this linear perturbation has a positive slope along both $Q_\parallel$ and $Q_\perp$. The transition state is displaced to larger $Q_\parallel$ values and to shorter $Q_\perp$ values. Hence a substitution that is energetically unfavorable on the reaction coordinate (i.e., makes it more difficult) increases its value at the transition state. The substitution that is also energetically unfavorable on the perpendicular coordinate (i.e., makes it more difficult) decreases its value at the transition state. The results are reversed if the substitutions are energetically favorable and facilitate the coordinates.

As an example, consider the change of bromine to chlorine in the substitution reaction

$$OH^- + CH_3Br \rightarrow CH_3OH + Br^- \tag{2-28}$$

Since a CCl bond is stronger than a CBr bond, the substitution is energetically unfavorable for both parallel coordinate

$$OH\rightarrow\ \leftarrow CH_3\!-\!Br\rightarrow \tag{2-29}$$
$$\text{(Cl)}$$

and perpendicular coordinate

$$\leftarrow OH\!-\!CH_3\!-\!Br\rightarrow \tag{2-30}$$
$$\text{(Cl)}$$

[see (2-26) and (2-27)]. The first coordinate should be more advanced for chlorine than for bromine and the second less, at the transition state. The conclusion is that the CO bond should be shorter in the presence of chlorine, whereas there is doubt

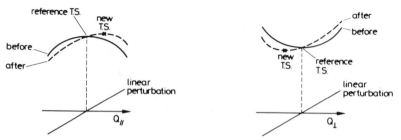

FIGURE 2-14.   Effect of adding a linear energy perturbation with *positive* slope to the parabolic potentials for $Q_\parallel$ and $Q_\perp$. Crosses represent new position of transition state (T.S.).

as to the outcome for the C-halogen bond.* Experimentally, the kinetic isotope effect for solvolysis of CH_3Cl in D_2O versus H_2O is indeed greater than that for solvolysis of CH_3Br,[63] indicating that OC bond making is more complete at the transition state for CH_3Cl. Attempts to remove the ambiguity of the model for the direction of change in length of the reacting bond which is to be broken have been made[64] for S_N2 transition states.

Arguments similar to Thornton's rules can be applied to the study of steric effects on transition states. If passage through the transition state corresponds to a decrease in steric strain, crowding in the reactant will cause the transition state to come earlier.† It has been pointed out[67] in the case of nucleophilic addition to ketones that increasing "congestion" also displaces the transition state toward starting materials. In fact, in the extreme case of a reactive center surrounded by distant bulky groups such that the energy needed by the reagent to pass by these groups is rate–determining, the transition state may have no bond-making character at all.

2.8 Hammond Postulate

In 1955 Hammond[68] postulated that "if two states, as for example, a transition state and an unstable intermediate, occur consecutively during a reaction process and have nearly the *same energy content,* their interconversion will involve only a *small reorganization of molecular structure.*" This postulate has been extremely useful in comparing the structure of transition states directly to those of reactants and products.

Take a highly exothermic reaction. Its energy profile along the reaction coordinate generally resembles Fig. 2-15a: both transition state and reactant have nearly

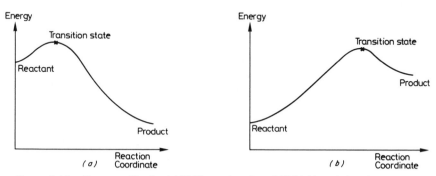

FIGURE 2-15.   Energy profiles for *(a)* highly exothermic and *(b)* highly endothermic reactions.

---

*In the latter case the terms "lengthening" and "shortening" must be specified since the halogen atom changes. A "lengthening" implies that at the transition state the C—Cl bond, relative to a *normal* CCl bond, will be longer than the C—Br bond was relative to a normal CBr bond.

†See Ref. 65; for a different conclusion, see Ref. 66.

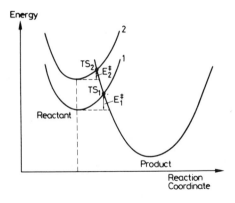

FIGURE 2-16. The potential-energy surface as a sum of two harmonic wells (exothermic reaction) (*TS* denotes the transition state). If the reaction is rendered more exothermic ($1 \rightarrow 2$), the transition state is moved closer to reactant ($TS_1 \rightarrow TS_2$), both in energy ($E_2^{\ddagger} < E_1^{\ddagger}$) and in position.

the same energy. But they are also close to each other on the reaction coordinate, in accordance with Hammond's prediction that the transition state will resemble the reactant. Similarly, the usual profile for a highly endothermic reaction (Fig. 2-15*b*) requires that transition state and product have nearly the same energy and accordingly correspond to neighboring values of the reaction coordinate; the transition state should resemble the product. For instance, in a solvolysis S_N1 reaction such as

$$C_2H_5X + OH^- \rightarrow C_2H_5{}^+ + X^- + OH^- \qquad (2\text{-}31)$$

to form a carbenium ion intermediate, the transition state is assumed to be so close to the ion that one correlates rates of solvolysis with the stability of the ion formed.

The Hammond postulate is really based on the idea that the form of the potential surface along the reaction coordinate can be approached by a sum of harmonic wells (Fig. 2-16). This idea has already been used in essence in interpreting the principle of least motion (Fig. 2-11). The transition state occurs at the intersection of the two parabola. If we make the reaction even more exothermic by lifting the energy of the reactant well, the transition state is moved even closer ($TS_1 \rightarrow TS_2$)

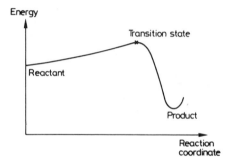

FIGURE 2-17. A non-Hammond energy profile.

to reactant, in both energy ($E_2^{\ddagger} < E_1^{\ddagger}$) *and* position. Hence similarity in energy does imply similarity in position.

Violations of the Hammond postulate require a special form of the potential surface, essentially one in which the energy profile stays *flat* over a large section of the reaction coordinate (Fig. 2-17); then closeness in energy for consecutive "states" (initial, final, or transition) does not imply closeness in geometry any more. Such flat profiles may occur if the force constant for one of the coordinates contributing to the reaction is very soft. A potential surface such as that shown in Fig. 2-17 might apply to the Diels-Alder reaction as written between *trans*-butadiene and ethylene:

$$(2\text{-}32)$$

The incipient phase of the reaction will be a rotation from transoid to cisoid form as the partners come together, with little cost; the transition state, much further along the path, will not resemble reactants. Caution should also be used in other cases, particularly when—as in E2 eliminations—the activated complex lies far above both reactants and products.[69]

2.9 Free-Energy Relationships

Free-energy relationships relate reaction rates to equilibrium constants for the equilibrium between reactants and products. Here we go one step further than the Hammond postulate since we attempt to relate the energy barrier to free-energy differences—approximated by the potential energy differences—between reactants and products. The existence of such relationships, however, is already obvious in the harmonic well treatment of the Hammond postulate (Fig. 2-16), for which higher exothermicity was shown to be related to lower activation energy.

Historically the first free-energy relationship to be discovered was the Brönsted relation.[70,*] Brönsted and Pedersen found that the rate $\mathbf{k}$ for acid catalysis (in which the catalyst donates a proton H^+ to the reacting system) is proportional to the strength of the acid catalyst as measured by its equilibrium constant for dissociation ($K = [X^-][H_3O^+]/[XH]$ for an acid XH):

$$\mathbf{k} = GK_{\text{catalyst}}^{\alpha} \qquad (2\text{-}33)$$

where α is an exponent smaller than unity and G is a constant for a series of similar

---

*According to Evans and Polanyi,[71] this discovery originated with Snethlage,[72] who first showed the parallelism between acid strength and catalytic molecular activity and with Taylor,[73] who first related them quantitatively.

catalysts with a fixed substrate. The constant G is a function of the reaction (and reaction conditions: solvent and temperature) whose rate **k** is being measured. A similar proportionality relation involving $K_{\text{substrate}}$ holds if we vary the substrate for fixed catalyst.

To interpret (2-33) in terms of potential-energy surfaces, we consider, with Bell and others,[74–77] the potential energy for general proton transfer

$$XH + Y \rightarrow X + HY \tag{2-34}$$

formed by two parabolic curves, just as in Fig. 2-16. Figure 2-18 shows these curves, as well as the displaced curve $(1 \rightarrow 2)$ for reactants if the energy difference between reactant and product is increased. The change in activation energy is given by

$$E_2^\ddagger - E_1^\ddagger = \Delta E - b \tag{2-35}$$

where b is the vertical distance separating the two transition structures TS_1 and TS_2. If curve 2 is replaced by its tangent at TS_2 and the product curve is assumed to be linear between TS_1 and TS_2 (Fig. 2-18, right-hand-side detail), it is possible to calculate b by evaluating the length $A \cdot TS_1$ successively as AB (in triangle $A \cdot B \cdot TS_2$) + $B \cdot TS_1$ (in triangle $B \cdot TS_1 \cdot TS_2$) and as $A \cdot TS_1$ in triangle $A \cdot C \cdot TS_1$:

$$b\left(\frac{1}{s_r} + \frac{1}{s_p}\right) = \frac{\Delta E}{s_r} \tag{2-36}$$

where s_r and s_p are the absolute values of the slopes, respectively, of the intersecting reactant and product curves; thus

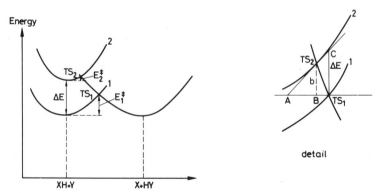

FIGURE 2-18.   Schematic potential-energy curves for proton transfer and relation between change in activation energy and ΔE. Length b is given by $B \cdot TS_2$ (see detail).

$$E_2^{\ddagger} - E_1^{\ddagger} = \Delta E \left(\frac{s_r}{s_r + s_p} \right) \qquad (2\text{-}37)$$

and the proportionality of activation energy to energy change in the reaction is established.

Even if one admits that ΔE is close to the enthalpy of reaction ΔH (neglecting zero-point vibrational energy differences between reactants and products, $\Delta E = \Delta H$ at $0°K$), the proportionality of the free-energy change ΔG to ΔH requires an "isokinetic" relationship of the type[78]

$$\delta \Delta H = \beta \delta \Delta S \qquad (2\text{-}38)$$

where ΔS is the entropy of reaction.

The most famous free-energy relationship is the Hammett equation[79,80] in which the effect of a substituent on the reactivity of a compound can be represented by the *product* of two factors, one of which (σ) is characteristic of substituent, the other (ρ) characterizes the reaction under consideration, in particular its sensitivity to substituent influences. This equation takes the general form

$$\log \frac{k}{k_0} = \rho \sigma \qquad (2\text{-}39)$$

where σ is proportional to the free-energy increment (assumed constant) brought about by the particular substituent that gives the rate k. The underlying interpretation, with the use of potential-energy surfaces, must be much the same as that of the Brönsted relation.

2.10 Competing Pathways and Curtin-Hammett Principle

Take a reactant that has several different conformers, such as an asymmetric ketone

$$(2\text{-}40)$$

and have it react, for instance, by reducing it with $NaBH_4$. If we wish to predict the product distribution, should we consider only the *most stable* conformer of reactant?

The answer was given by Curtin and Hammett[81–83]: if the conformers react slowly with respect to their interconversion rate, then the product distribution is related to the difference in energy between the competing transition states. The result is the same as if there were a single starting material with two available pathways. Schematically:

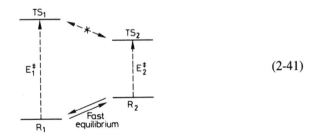

$$(2\text{-}41)$$

The essential criterion is the barrier height for reactant interconversion (5.5 kJ/mole for rotation around C_1C_2 in the ketone) versus barrier height for reaction (30–40 kJ/mole for reduction). A further parameter is the barrier between TS_1 and TS_2:

1.   If it is *high* relative to the internal energy of reactants in that region, there will be two distinctly accessible cols [see (2-41) and Fig. 2-19a]. Since the collisional energy required to activate either is much larger than their initial interconversion energy, *both* reactants can go through either col. For R_1 to go through TS_2, it need only borrow first the interconversion coordinate $R_1 \rightarrow R_2$. The product distribution is

$$\frac{P_1}{P_2} = \frac{\exp(-E_1^{\ddagger}/kT)}{\exp(-E_2^{\ddagger}/kT)} \qquad (2\text{-}42)$$

which is the normal expression of the Curtin-Hammett principle.* It means that the product reached from the most reactive conformer is selected in priority even if this conformer is less abundant.

2.   If it is small, there is only one effective col (Fig. 2-19b) and all paths will go through TS_2. Indeed, R_1, instead of going straight up to TS_1, will turn around along the conformational coordinate direction and go through TS_2. The exclusive product should be P_2.

H	O	H	O	Me	O	Me	O

$$\mathbf{d}\,(-30°) \qquad \mathbf{a}\,(0°) \qquad \mathbf{b}\,(120°) \qquad \mathbf{c}\,(150°)$$

←————————————Low-energy interconversion————————————→

$$(2\text{-}43)$$

*Slanina[84] makes a general study of the relation between activation parameters of partial rate processes and the rate characteristics of the overall process; see also Ref. 85.

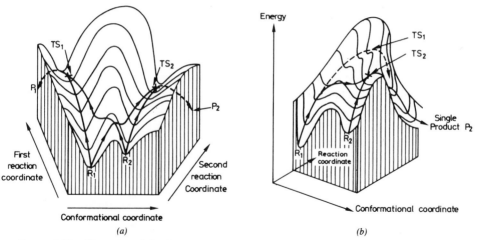

FIGURE 2-19.   Illustration of the Curtin-Hammett principle, in which the unstable conformer corresponds to the more stable transition state. *(a)* Barrier between transition states: the diagonal paths $R_1 TS_2$ and $R_2 TS_1$ can be regarded as interconversions followed by least-energy pathways. *(b)* No barrier between transition states: TS_1 would be reacted only if R_1 did not use the conformational coordinate to find its least energy path through TS_2.

As an illustration, consider the chloropropanol molecule and four of its ground state conformations in (2-43). The two more stable ones are **a** and **b**, but **c** and **d** are higher by only roughly 4 kJ/mole. In the presence of an approaching H⁻ ion, transition states can be formed for each conformer, which correspond to two distinct diasteroisomeric pathways, which we denote by unprimed and primed letters:

$$(2\text{-}44)$$

TS_d TS_a TS'_b TS'_c

(TS_b, TS_c much higher)   (TS'_a, TS'_d much higher)

←————————————No interconversion————————————→

The two stable transition-state conformations are TS'_c and TS_d, as predicted by Chérest et al.[86] and demonstrated by Anh and Eisenstein.[87]

Application of the Curtin-Hammett principle to this reaction requires exclusive consideration of the two cols TS'_c and TS_d. The two cols are on entirely distinct paths, with no passage possible between them (see Fig. 2–19a; internal rotation does not convert TS_c' into TS_d: the hydrogen atom would have to migrate around the molecule). Hence the product ratio for the two diastereoisomers depends exclusively on the ratio of TS_c' and TS_d, regardless of the relative (but not too different) energies of **a, b, c,** and **d.** Going one step further, and considering the approach of H^- exclusively on the front side, say, we now compare the behavior of conformations **a** and **d.** Here there is no barrier[87] between TS_a and TS_d, even though the former is higher in energy. As stated previously, both conformers then go through the same transition state TS_d and give the same product. The situation is similar to Fig. 2–19b, although **d,** contrary to R_2, is not a secondary minimum but simply lies above **a,** on the ground-state conformational curve.

Returning to Fig. 2–19a, we also have an illustration of *kinetic* versus *thermodynamic* control.* The forward reaction gives noninterconvertible products P_1 and P_2 whose ratio is determined by the heights of the cols: it is kinetically controlled. The *backward* reaction, however, despite the favored path through TS_2, will lead to products R_1 and R_2 in ratios depending on their relative energies: thermodynamic control dominates because R_1 and R_2 are interconverting speedily relative to the actual rate of reactions $P_1 \to TS_1$ and $P_2 \to TS_2$.

Although the nature of the control is different in the forward and the backward reactions, the "principle of microscopic reversibility"[89] requires that the relative number of molecules choosing the path through TS_1 and that through TS_2 be the same regardless of the direction of the reaction.

In certain cases kinetic control may lead to products in ratios resembling (but different from) those predicted from thermodynamic control. This occurs for "late" transition states, with an interconversional barrier resembling that for the final product conformers. Since $E_2^{\ddagger} - E_1^{\ddagger}$ is close to $P_2 - P_1$, *"product development control"* occurs.[90]

References

1.   S. D. Peyerimhoff and R. J. Buenker, *Chem. Phys. Lett.* **16,** 235 (1972).
2.   G. Herzberg, *Electronic Spectra and Electronic Structure of Polyatomic Molecules* (Van Nostrand, New York, 1966), Fig. 166.
3.   S. D. Peyerimhoff and R. J. Buenker in *The New World of Quantum Chemistry,* edited by B. Pullman and R. Parr (Reidel, Dordrecht, 1976), p. 213.
4.   P. G. Mezey in *Applications of MO Theory in Organic Chemistry,* edited by I. G. Csizmadia (Elsevier, Amsterdam, 1977), p. 127 and Fig. 17.
5.   A. H. Andrist, *J. Org. Chem.* **38,** 1772 (1973).
6.   R. Hoffmann, S. Swaminathan, B. G. Odell, and R. Gleiter, *J. Am. Chem. Soc.* **92,** 7091 (1970).
7.   S. Winstein, E. Clippinger, A. H. Fainberg, and G. C. Robinson, *J. Am. Chem. Soc.* **76,** 2597 (1954); M. Szwarc, *Acc. Chem. Res.* **2,** 87 (1969).
8.   N. L. Owen in *Internal Rotation in Molecules,* edited by W. J. Orville-Thomas (Wiley, New York, 1973) Section 6.15.
9.   J. E. Baldwin, A. H. Andrist and R. K. Pinschmidt, *Acc. Chem. Res.* **5,** 402 (1972).
10.   J. P. Lowe, *J. Chem. Ed.* **51,** 785 (1974).

*For a discussion of thermodynamic versus kinetic control, see Ref. 88.

11. S. Arrhenius, *Z. Physik. Chem. (Leipzig)* **4**, 226 (1889); K. J. Laidler, *Chemical Kinetics* (McGraw-Hill, New York, 1965), p. 50.
12. Y. Jean, *Nouv. J. Chim.* **4**, 11 (1980).
13. H. Eyring, *J. Chem. Phys.* **3**, 107 (1935).
14. M. G. Evans and M. Polanyi, *Transact. Faraday Soc.* **31**, 875 (1935).
15. K. J. Laidler, *Chemical Kinetics* (McGraw-Hill, New York, 1965) pp. 85 sqq.
16. M. Menzinger and R. Wolfgang, *Angew. Chem. Internatl. Ed.* **8**, 438 (1969).
17. W. F. K. Wynne-Jones and H. Eyring, *J. Chem. Phys.* **3**, 492 (1935).
18. R. A. Grieger and C. A. Eckert, *J. Am. Chem. Soc.* **92**, 7149 (1970).
19. C. A. Stewart, *J. Am. Chem. Soc.* **94**, 635 (1972).
20. P. Kohnstamm and F. E. C. Scheffer, *Proc. Roy. Acad. Amsterdam (Koninklijke Akademie van Wetenschappen te Amsterdam)* **13**, 789 (1910).
21. R. Marcelin, *Comptes Rendus Acad. Sci. (Paris)* **157**, 1052 (1910); *Annales de Physique* **3**, 120 (1915).
22. F. K. Fong, *J. Am. Chem. Soc.* **96**, 7645 (1974).
23. J. N. Murrell and K. J. Laidler, *Transact. Faraday Soc.* **64**, 371 (1968).
24. R. M. Noyes, *J. Chem. Phys.* **48**, 323 (1968); M. R. Wright and P. G. Wright, *J. Phys. Chem.* **74**, 4394 (1970); P. G. Wright and M. R. Wright, ibid. **74**, 4398 (1970).
25. J. N. Murrell in *Quantum Theory of Chemical Reactions,* Vol. **1**, edited by R. Daudel, A. Pullman, L. Salem, and A. Veillard (Reidel, Dordrecht, 1980), p. 161; J. N. Murrell and S. Farantos, *Molec. Phys.* **34**, 1185 (1977); J. N. Murrell, *J. Chem. Soc. Chem. Commun.* 1044 (1972).
26. J. N. Murrell and G. L. Pratt, *Transact. Faraday Soc.* **65**, 1680 (1969); K. J. Laidler in *Reaction Transition States,* edited by J. E. Dubois (Gordon and Breach, London, 1970), p. 23.
27. E. Pollak and P. Pechukas, *J. Am. Chem. Soc.* **100**, 2984 (1978).
28. R. E. Stanton and J. W. McIver, *J. Am. Chem. Soc.* **97**, 3632 (1975); J. W. McIver, *Acc. Chem. Res.* **7**, 72 (1974).
29. P. Pechukas, *J. Chem. Phys.* **64**, 1516 (1976).
30. K. Muller, *Angew. Chem. Internatl. Ed.* **19**, 1 (1980).
31. D. J. Wilde, *Optimum Seeking Methods* (Prentice-Hall, Englewood Cliffs, N.J., 1964).
32. E. A. McCullough and D. M. Silver, *J. Chem. Phys.* **62**, 4050 (1975).
33. N. M. Witriol, J. D. Stettler, M. A. Ratner, J. R. Sabin, and S. B. Trickey, *J. Chem. Phys.* **66**, 1141 (1977).
34. K. Fukui, *J. Phys. Chem.* **74**, 4161 (1970); K. Fukui, *J. Am. Chem. Soc.* **97**, 1 (1975), footnote 13; K. Fukui, *Acc. Chem. Res.* **14**, 353 (1981).
35. P. Mathias and W. A. Sanders, *J. Chem. Phys.* **64**, 388 (1976).
36. J. C. Polanyi, *Acc. Chem. Res.* **5**, 161 (1972).
37. N. H. Hijazi and K. J. Laidler, *J. Chem. Phys.* **58**, 349 (1973).
38. P. Pulay in *Applications of Electronic Structure Theory,* edited by H. F. Schaefer (Plenum, New York, 1977), Chapter 4.
39. A. Komornicki, K. Ishida, K. Morokuma, D. Ditchfield, and M. Conrad, *Chem. Phys. Lett.* **45**, 595 (1977).
40. C. Leforestier, *J. Chem. Phys.* **68**, 4406 (1978).
41. J. D. Goddard, N. C. Handy, and H. F. Schaefer, *J. Chem. Phys.* **71**, 1525 (1979).
42. J. A. Pople, R. Krishnan, H. B. Schlegel, and J. S. Binkley, *Internatl. J. Quant. Chem.,* Proceedings 13th International Symposium Quantum Chemistry (Interscience, New York, 1979) p. 226.
43. C. Leforestier D.Sc. thesis, Université de Paris-Sud, Orsay, France, 1979, p. 6.
44. J. W. McIver and A. Kormonicki, *J. Am. Chem. Soc.* **94**, 2625 (1972).
45. D. Poppinger, *Chem. Phys. Lett.* **35**, 550 (1975).
46. Y. Jean, L. Salem, J. S. Wright, J. A. Horsley, C. Moser, and R. M. Stevens, *Pure Appl. Chem. Suppl.* **1**, 197 (1971); L. Salem, *Acc. Chem. Res.* **4**, 322 (1971).
47. K. Ishida, K. Morokuma, and A. Komornicki, *J. Chem. Phys.* **66**, 2153 (1977).
48. M. J. S. Dewar and S. Kirschner, *J. Am. Chem. Soc.* **93**, 4291 (1971).

49. A. Komornicki and J. W. McIver, *J. Am. Chem. Soc.* **96**, 5798 (1974).
50. T. A. Halgren and W. N. Lipscomb, *Chem. Phys. Lett.* **49**, 225 (1977).
51. J. Pancir, *Coll. Czech. Chem. Commun.* **40**, 1112 (1975).
52. K. Müller and L. D. Brown, *Theor. Chim. Acta (Berl.)* **53**, 75 (1979).
53. J.-A. Muller, *Bull. Soc. Chim. Fr.* **45**, 438 (1886); E. Peytral, ibid. **29**, 39 (1921).
54. J.-A. Muller and E. Peytral, *Comptes Rendus Acad. Sci.* **179**, 831 (1924).
55. F. O. Rice and E. Teller, *J. Chem. Phys.* **6**, 489 (1938); ibid. **7**, 199 (1939); J. Franck and E. Rabinowitch, *Z. Elektrochem.* **36**, 794 (1930).
56. J. Hine, *J. Org. Chem.* **31**, 1236 (1966).
57. A. J. Birch, A. L. Hinde, and L. Radom, *J. Am. Chem. Soc.* **102**, 3370 (1980).
58. O. S. Tee, *J. Am. Chem. Soc.* **91**, 7144 (1969); O. S. Tee and K. Yates, ibid. **94**, 3074 (1972); O. S. Tee, J. A. Altmann, and K. Yates, ibid. **96**, 3141 (1974).
59. S. Ehrenson in *Chemical and Biochemical Reactivity,* edited by B. Pullman (Israel Academy of Sciences and Humanities, Jerusalem, 1974), p. 113.
60. S. Ehrenson, *J. Am. Chem. Soc.* **98**, 6081 (1976).
61. I. G. Csizmadia, G. Theodorakopoulous, H. B. Schlegel, M. H. Whangbo, and S. Wolfe, *Can. J. Chem.* **55**, 986 (1977).
62. E. R. Thornton, *J. Am. Chem. Soc.* **89**, 2915 (1967).
63. C. G. Swain and E. R. Thornton, *J. Am. Chem. Soc.* **84**, 822 (1962).
64. J. C. Harris and J. L. Kurz, *J. Am. Chem. Soc.* **92**, 349 (1970).
65. C. G. Swain and N. D. Hershey, *J. Am. Chem. Soc.* **94**, 1901 (1972).
66. W. J. le Noble and T. Asano, *J. Am. Chem. Soc.* **97**, 1778 (1975).
67. W. T. Wipke and P. Gund, *J. Am. Chem. Soc.,* **98**, 8107 (1976).
68. G. S. Hammond, *J. Am. Chem. Soc.* **77**, 334 (1955).
69. D. Farcasiu, *J. Chem. Ed.* **52**, 76 (1975).
70. J. N. Brönsted and K. J. Pedersen, *Z. Physik. Chem. (Leipzig)* **108**, 185 (1924).
71. M. G. Evans and M. Polanyi, *Transact. Faraday Soc.* **32**, 1333 (1936).
72. H. C. S. Snethlage, *Z. Elektrochem.* **18**, 539 (1912).
73. H. S. Taylor, *Z. Elektrochem.* **20**, 201 (1914).
74. J. Horiuti and M. Polanyi, *Acta Physicochim. URSS* **2**, 505 (1935).
75. R. P. Bell, *Proc. Roy. Soc.* **A154**, 414 (1936); R. P. Bell, *The Proton in Chemistry* (Methuen, London, 1959), Chapter 10.
76. M. G. Evans and M. Polanyi, *Transact. Faraday Soc.* **34**, 11 (1938).
77. M. J. S. Dewar, *The Molecular Orbital Theory of Organic Chemistry* (McGraw-Hill, New York, 1968), Section 8.3.
78. O. Exner, *Progr. Phys. Org. Chem.* **10**, 411 (1973).
79. L. P. Hammett, *Chem. Rev.* **17**, 125 (1935); *J. Am. Chem. Soc.* **59**, 96 (1937); *Transact. Faraday Soc.* **34**, 156 (1938).
80. R. W. Taft, *J. Am. Chem. Soc.* **74**, 2729, 3120 (1952).
81. D. Y. Curtin, *Rec. Chem. Progr.* **15**, 111 (1954).
82. E. Eliel, *Stereochemistry of Carbon Compounds* (McGraw-Hill, New York, 1962), p. 151.
83. H. Kagan, *Organic Stereochemistry* (Edward Arnold, London, 1979), p. 134.
84. Z. Slanina, *Coll. Czech. Commun.* **42**, 1914 (1977).
85. J. I. Seeman and W. A. Farone, *J. Org. Chem.* **43**, 1854 (1978); N. S. Zefirov, *Tetrahedron* **33**, 2719 (1977).
86. M. Chérest, H. Felkin, and N. Prudent, *Tetrahedron Lett.* 2201 (1968).
87. N. T. Anh and O. Eisenstein, *Nouv. J. Chimie* **1**, 61 (1977).
88. R. Bentley, *Molecular Asymmetry in Biology,* Vol. I (Academic, New York, 1969), p. 75.
89. O. K. Rice, *Statistical Mechanics, Thermodynamics and Kinetics* (Freeman, San Francisco, 1967), Chapter 17.
90. W. G. Dauben, G. F. Fonken, and D. S. Noyce, *J. Am. Chem. Soc.* **78**, 2579 (1956).

3

Electronic Structure of Reactive Species and Important Intermediates

A prerequisite for determining reaction pathways is a good understanding of the electronic structure of possible reaction intermediates. By "electronic structure" we mean both ground and excited electronic configurations of the system if we use molecular orbital theory, or if we use valence-bond theory, the appropriate resonance forms. In either case the proper states have to be determined as appropriately weighted combinations of configurations or of resonance forms.

For most closed-shell molecules, the description is straightforward: the equilibrium geometry corresponds to a ground singlet state, and the excited states are well separated from the latter. In certain relatively unstable systems, such as carbenium ions, however, there may be interesting conformational problems involving neighboring valence structures.

For "open-shell" molecules, however, the situation is different. By "open-shell" we mean that there are more low-lying valence atomic or hybrid orbitals available to the electrons than there are total number of available electron pairs. In methylene, for instance, two carbon electrons can choose between two orbitals (one p and one hybrid sp^2) on carbon. In the oxygen molecule two electrons can again choose between two antibonding π-type molecular orbitals built on the two oxygen atoms. For such systems, there are a number of relatively low–lying states, which we construct. The following molecules are studied here:

:CH$_2$	Methylene and cyclic derivatives thereof		
·O=O·	Oxygen	$\dot{C}H_2$—$\dot{C}H_2$	Twisted ethylene (Dimethylene)

$$
\begin{array}{ll}
\overset{\displaystyle O}{\underset{\displaystyle \cdot O \qquad O\cdot}{\diagup \diagdown}} & \text{Ozone} \\
\end{array}
$$

$$
\begin{array}{ll}
\overset{\displaystyle CH}{\underset{\displaystyle \dot{C}H_2 \qquad CH_2}{\diagup \diagdown\!\!\!=}} & \text{Allyl} \\
\end{array}
$$

$$
\begin{array}{ll}
\overset{\displaystyle CH_2}{\underset{\displaystyle \dot{C}H_2 \qquad \dot{C}H_2}{\diagup \diagdown}} & \text{Trimethylene} \\
\end{array}
$$

$$
\begin{array}{ll}
\overset{\displaystyle CH}{\underset{\displaystyle :CH \qquad CH_2}{\diagup \diagdown\!\!\!=}} & \text{Vinylmethylene} \\
\end{array}
\qquad (3\text{-}1)
$$

$$
\begin{array}{ll}
\begin{array}{l} CH\!-\!CH \\ |\ \ ..\ \ | \\ CH\!-\!CH \end{array} & \text{Cyclobutadiene} \\
\end{array}
$$

$$
\begin{array}{ll}
\overset{\displaystyle CH_2}{\overset{\displaystyle \|}{\underset{\displaystyle \dot{C}H_2 \qquad \dot{C}H_2}{\underset{C}{\diagup \diagdown}}}} & \text{Trimethylenemethane} \\
\end{array}
$$

With the exception of allyl, these molecules belong to the general family of molecules called *diradicals,* defined by Berson[1] as "even-electron molecules that have one bond less than the number permitted by the standard rules of valence."

3.1 Diradicals and Zwitterions and the Four Diradical and Zwitterionic States

Consider the hydrocarbon molecule and the nitrile oxide

$$R-C\!\equiv\!\overset{\oplus}{N}\!-\!\overset{\ominus}{O} \tag{3-2}$$

The former ("Schlenk's biradical") is commonly considered to be a triplet diradical[2] and the latter, a zwitterion or "1,3 dipole".[3] However, a closer look shows that this description must be qualified. We can imagine distributions of the electrons that lead to a different description of each molecule. For the hydrocarbon, in addition to the two conventional descriptions with paired or unpaired odd spins, there are two resonance structures in which one carbon atom is negative and carries the electron pair while the other terminus is positive. Those structures can form an in-phase and an out-of-phase combination [see (3-3)]. For the nitrile oxide, we can break up the electron pair on oxygen and form two unpaired electron structures (parallel or antiparallel spin). It is even possible, by depleting the oxygen atom entirely and making a negative carbon atom, to obtain a zwitterionic description with reversed polarity [see (3-4)].*

$$\text{(3-3)}$$

$$R-C\!\equiv\!\overset{\oplus}{N}\!-\!\overset{\ominus}{\overset{..}{O}} \qquad R-\overset{\uparrow}{C}\!=\!\overset{\uparrow}{N}\!-\!O \qquad R-\overset{\uparrow}{C}\!=\!\overset{\downarrow}{N}\!-\!O \qquad R-\overset{\ominus}{\overset{..}{C}}\!=\!\overset{\oplus}{N}\!=\!O \tag{3-4}$$

To the four different descriptions of Schlenk's biradical and of the 1,3 dipole correspond *four* different electronic states.[4,5] Furthermore, although their ordering is different in the diradical and the zwitterion, the same four types of states occur in both systems: (1) a triplet "diradical" state with two unpaired electrons of

---

*There is a second resonance contributor to the ground-state nitrile oxide (3-2), which is $R-\overset{\oplus}{C}\!=\!\overset{\ominus}{N}\!-\!O$. See also Section 6.3.

parallel spin; (2) a singlet "diradical" state with two unpaired electrons but anti-parallel spin; (3) a "zwitterionic" or "ion-pair" singlet state with charge polarized in one direction; and (4) a "zwitterionic" or "ion-pair" singlet state with the opposite polarization (or as an alternative to the last two states, a pair of zwitterionic states with respectively in-phase and out-of-phase resonance between oppositely polarized structures; see Section 1.3).

The quantum-mechanical description of these four states is straightforward.[4] Two limiting cases are of interest. In the first the two odd available orbitals for the unpaired electrons, although not necessarily localized, belong to different symmetry representations of the molecular point group ("heterosymmetric" diradical). This is true for the nitrile oxide where these orbitals are two orthogonal π orbitals[6,*] (the molecular axis is the z axis)

$$a = 0.68\,\phi_C^x - 0.67\,\phi_N^x + 0.30\,\phi_O^x \quad \text{(higher)}$$
$$b = 0.56\,\phi_C^y + 0.21\,\phi_N^y - 0.80\,\phi_O^y \quad \text{(lower)} \tag{3-5}$$

The four states are then described as follows:

^3ab (diradical triplet: one electron on oxygen; carbon and nitrogen compete for the other)

^1ab (diradical singlet: one electron on oxygen; carbon and nitrogen compete for the other) $\tag{3-6}$

$^1b^2 \underset{(\lambda > 0)}{-} \lambda^1a^2$ (lower zwitterionic singlet; electron pair essentially on oxygen; hole on nitrogen)

$^1a^2 \underset{(\mu > 0)}{+} \mu^1b^2$ (higher zwitterionic singlet; electron pair shared by carbon and nitrogen; hole on oxygen)

The low weight of $^1a^2$ on oxygen indicates that the last state of high energy may in fact be better described by the resonance structure $R—\overset{\ominus}{C}=N—\overset{\oplus}{O}$ than by $R—\overset{\ominus}{\ddot{C}}=N=\overset{\oplus}{O}$ of (3-4). Similarly, the electron distribution given by configuration ab indicates that a single resonance structure, as in (3-4), does not suffice to describe the diradical states:

$$R—\overset{\uparrow}{C}=N—\overset{\uparrow}{O} \longleftrightarrow R—\overset{\ominus}{\ddot{C}}=\overset{\oplus}{N}—\overset{\uparrow}{O} \tag{3-7}$$

---

*Compare with Ref. 7, where orbital a appears to be in error. The author is grateful to P. Hiberty for a discussion of resonance in the nitrile oxides.

Equations (3-5)–(3-7) give us a foretaste of the advantages obtained by the interplay of molecular orbital theory and resonance theory:

The molecular orbital configurations indicate the appropriate resonance structures that must intervene.

They also indicate the *planes* of the odd electrons: the odd electrons on oxygen and nitrogen are in orthogonal planes, whereas the odd spin density on carbon arises from delocalization in both planes (starting at oxygen in one plane and at nitrogen in the other).

In turn, the resonance structures, which now number 2, indicate that there must be a *second* triplet diradical state (and also a second singlet diradical state) of not too high energy!

In the other limiting case (Schlenck's biradical) the odd orbitals a and b are symmetry-equivalent, being related by a symmetry element, axis, or plane of the molecular point group ("homosymmetric" diradicals). The proper starting molecular orbitals are then $a + b$ and $a - b$, and the electronic configurations for the four states are given by[4,5]

Energy

$^3(a + b)(a - b)$: diradical triplet $\qquad\qquad J_{ab} - K_{ab}$

$^1(a + b)^2 - {}^1(a - b)^2$: diradical singlet $\qquad J_{ab} + K_{ab}$

$^1(a + b)(a - b)$: out-of-phase zwitterionic singlet $\qquad \dfrac{1}{2}(J_{aa} + J_{bb}) - K_{ab}$

$^1(a + b)^2 + {}^1(a - b)^2$: in-phase zwitterionic singlet $\qquad \dfrac{1}{2}(J_{aa} + J_{bb}) + K_{ab}$

$$(3\text{-}8)$$

In homosymmetric diradicals the energies of the four states—relative to the common energy of a and b—are given in simple form in terms of J_{ab}, the Coulomb repulsion integral between a and b; J_{aa}, the self-repulsion integral for a; and K_{ab}, the exchange integral between a and b (see Chapter 1, footnote a in Table 1-2). The terminologies "*out-of-phase*" and "*in-phase*" for the zwitterionic singlets correspond identically with those used in Section 1.3; the proper wave functions for (3-8) are identical with those for (1-10).

In diradical–like or zwitterionic–like systems without any symmetry the triplet state is still given by the single configuration ^3ab, where a and b are the odd orbitals. However, the three singlet states are all mixtures of three configurations

$$S_1, S_2, S_3 \equiv \lambda(^1a^2) + \mu(^1b^2) + \nu(^1ab) \qquad (3\text{-}9)$$

From the purely theoretical point of view, molecules that behave chemically as diradicals also have intriguing quantum-mechanical properties:

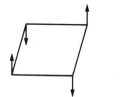

FIGURE 3-1.   Spin density waves in cyclobutadiene[13] (see Section 7.4). The arrows represent *directions of unpaired spin* and should not be construed as full unpaired electrons. *(left)* Singlet diradical state. *(right)* Triplet diradical state.

1.   A restricted closed-shell molecular orbital wave function such as (1-18) becomes "unstable" to the splitting of the orbitals into two groups and the formation of a more stable open-shell unrestricted solution, with triplet character, of type (1-22). The study of such "triplet instability"[8],* in diradicals is associated with the names of Fukutome, Yamaguchi, Koutecky, and Bonacic-Koutecky.[10-12]

2.   Associated with this triplet-unstable solution are the so-called spin-density waves"[13],† (see Section 7.4), which correspond to real spin-density distributions in which the two unpaired electrons occupy distinct spatial orbitals. For a stretched H_2 "diradical," these spatially distinct orbitals are similar to the GVB orbitals shown in Fig. 1-3; for cyclobutadiene,[13] they are shown in Fig. 3-1.

3.   Molecular systems may be defined as being diradicals according to the existence or nonexistence of such spin-density waves[11,13] or related properties.[15]

3.2   Methylene, Cyclic Ylidenes and Isoelectronic Molecules

The methylene molecule has been an important system in revealing the predictive power of quantum chemistry. Figure 3-2 gives the energy of the ground triplet state of methylene as a function of $\widehat{HCH}$ angle and CH bond length from the famous work in which Schaefer and Bender[16] proved the triplet to be bent rather than linear, as the experimental results then seemed to indicate. Figure 3-3 correlates the states of bent methylene with those of linear methylene.[16,17]

In bent CH_2 the two available orbitals that are not totally filled are an sp^2-like hybrid and a p-type orbital perpendicular to the molecular plane. The states of this "heterosymmetric" diradical can thus be written directly from (3-6). Their energies, geometries, and appropriate resonance formulation are given in Table 3-1. The resonance scheme for the lower singlet reveals immediately its "pair-hole" or zwitterionic character. The ground triplet, on the other hand, is a true diradical.

*Löwdin[9] first stated that an approximate trial wave function, corresponding to the absolute energy minimum, is not necessarily symmetry-adapted. The condition for triplet instability is $^3E_{a \to b} - K_{ab} < 0$, where a is the top (doubly) occupied orbital in the restricted determinant and b is the lowest unoccupied orbital, $^3E_{a \to b}$ is the corresponding triplet excitation energy for a one-electron transition from a to b, and K_{ab} is the exchange integral between a and b.

†For similar waves in monoradicals, see Ref. 14.

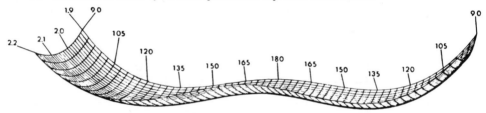

FIGURE 3-2.   Theoretical potential-energy surface for the $^3B_1$ ground state of CH_2. The total energy is plotted as a function of bond distance (in a.u.) and bond angle.[16] Reprinted with permission from S. V. O'Neil, H. F. Schaefer, and C. F. Bender, *J. Chem. Phys.* **55**, 162 (1971). Copyright 1971 by American Institute of Physics.

A great deal of controversy has surrounded the exact size of the singlet-triplet energy gap, which is now thought to be in the region of 43–48 kJ/mole.[19–22]

(3-10)

Substitution of the methylene hydrogens will affect the relative stability of lowest triplet and singlet. For instance, substitution of one hydrogen by fluorine lowers the singlet state by some 90 kJ/mole, thus rendering it the ground state, whereas a second fluorine atom stabilizes the singlet by an additional 150 kJ/mole.[23] The strong σ-type attractive interaction of the fluorines, because of their high electronegativity, favors the zwitterionic singlet[23,24] through structures of the type

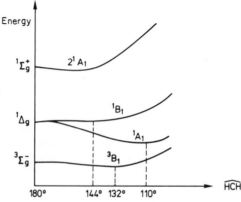

FIGURE 3-3. Qualitative behavior of the potential energy surfaces for the lowest states of methylene.[16,17]

TABLE 3-1
Calculated Properties of Methylene (CH$_2$)^a

State (C_{2v} Symmetry)	Energy		Geometry		Resonance Formulation	
	Absolute (a.u.)	Relative (eV)	$\widehat{HCH}$ (deg.)	R_{CH} (Å)	Major Structure	Minor Structure
2^1A_1	—	—	—	—	($^1\pi^2$)	($^1\sigma^2$)
$^1B_1$	[−38.9114]	[1.94]	143.8	1.092		($^1\sigma\pi$)
$^1A_1$	−39.0434	0.47	101.4 (102.4)	1.112 (1.11)	($^1\sigma^2$)	($^1\pi^2$)
$^3B_1$	−39.0607	0	132.3 (134)	1.083 (1.08)		($^3\sigma\pi$)

Source: Refs. 16, 18, 20.

^{a}(Experimental values in parentheses). Method: for the $^1A_1$ state, multiconfiguration (MC) SCF orbitals used as basis molecular orbitals, followed by configuration interaction performed with all singly and doubly replaced configurations with respect to the two reference ($^1\sigma^2$, $^1\pi^2$) configurations; for the $^3B_1$ state, a starting UHF determinant followed by similar configuration interaction; atomic orbitals are contracted Gaussian basis set, including carbon 3d orbitals. Results for the two lowest states calculated from Ref. 20 and third state, from Ref. 16. The relative energy of the third state compares this state to ground state with a similar methodology. For experimental values for the two lowest singlet states, see Ref. 18.

$$(3\text{-}11)$$

Back donation from the π lone pair of fluorine into the empty $p\pi$ orbital of carbon also contributes to the singlet[25,26] and to a lesser degree to the triplet[24] where this orbital is already partially occupied; this appears to be the dominant effect.[26]

$$(3\text{-}12)$$

For lithium and cyano substitution, π donation and electronegativity proceed in the opposite direction[23,24] and both put the triplet back below the singlet (where no π back donation from substituent into carbon is possible).

A crucial factor is the energy separation between σ and π orbitals on carbon. If σ is very low, the singlet wins by pairing electrons in the orbital; as π falls down, the electrons can go into two nearly equienergetic orbitals if their spins are parallel. The triplet becomes more stable and the singlet tends to become linear ($^1\Delta_g$) with equal weights of the two configurations $^1\sigma^2$ and $^1\pi^2$.

Inclusion of the carbenoid atom into a conjugated ring can also influence the relative stability of triplet and of $^1\sigma^2$ singlet—and even that of the $^1\pi^2$ singlet. Hence, whereas cycloheptatrienylidene is fairly well represented[27] by

$$(3\text{-}13)$$

corresponding to the single $^1\sigma^2$ component, cyclopentadienylidene should tend to be

$$(3\text{-}14)$$

favoring the other, $^1\pi^2$, component. Its ground state would be expected to be a singlet, with a large contribution from this component.[28,*] Note that this state really

---

*For a review of the electronic structure and reactivity of cycloalkenecarbenes, see Ref. 29.

correlates with the *third* singlet (2^1A_1) of methylene. Finally, cyclopropenylidene

$$\triangleright: \qquad (3\text{-}15)$$

has a singlet ground state[25,30] with an electronic distribution similar to (3-13).

Finally the calculated properties of silylene[31] SiH_2 and of the nitrenium ion[32] NH_2^+, isoelectronic with methylene, are given in Table 3-2. The more electropositive nature of silicon relative to carbon and the more electronegative character of N^+, explain the singlet stabilization on the one hand [see (3-11)] and the triplet stabilization on the other [see (3-12) with no compensating double-bonded structure such as the left-hand one possible for NH_2^+].

3.3 Oxygen

The oxygen molecule or "dioxygen" is a diradical *par excellence*. The association of its paramagnetism with a triplet ground state dates back to 1929.[34,*] We consider, in turn, the molecular orbital description[36,37] and the valence–bond description[38] of O_2.

The molecular orbital description of the electronic states of O_2 in the literature has been marred by confusion between the use of *imaginary* orbitals—proper to the isolated molecule with $D_{\infty h}$ symmetry—and that of *real* orbitals, proper to the molecule in the presence of any external perturber, such as a reactant. A single-configuration description in one case may correspond to a mixture in the other.

Let π_x^* and π_y^* be the antibonding π molecular orbitals defined by

$$\pi_x^* = \pi_x^A - \pi_x^B$$
$$\pi_y^* = \pi_y^A - \pi_y^B \qquad (3\text{-}16)$$

and as shown in Fig. 3-4. We now construct the imaginary orbitals

$$\pi_+^* = \pi_x^* + i\pi_y^*$$
$$\pi_-^* = \pi_x^* - i\pi_y^* \qquad (3\text{-}17)$$

An electron in π_+^* has one unit of orbital angular momentum around the molecular axis in the $x \rightarrow y$ (anticlockwise) direction, whereas an electron in π_-^* has a unit of orbital angular momentum in the opposite (clockwise) direction. If the two odd electrons are paired in either π_+^* or π_-^*, the molecule has two units of angular momentum: the two corresponding states are the two components of a $^1\Delta_g$ state. If one electron occupies π_+^* and the other π_-^*, their angular momenta cancel each

---

*Previous discussions are due to Lewis.[35]

TABLE 3-2

Calculated Properties of Silylene (SiH_2) and Nitrenium Ion (N^+H_2)[a]

	Energy		Geometry			Energy		Geometry	
State (C_{2v})	Absolute (a.u.)	Relative (eV)	$\widehat{HSiH}$ (deg.)	r_{SiH} (Å)	State (C_{2v})	Absolute (a.u.)	Relative (eV)	$\widehat{HNH}$ (deg.)	r_{NH} (Å)
2^1A_1	—	—	—	—	2^1A_1	—	—	—	—
$^1B_1$	−289.9644	2.30	123.5	1.468	$^1B_1$	—	—	—	—
$^3B_1$	−290.0193	0.81	117.6	1.471	$^1A_1$	−55.1833	1.26	108.2	1.033
$^1A_1$	−290.0489	0	94.3	1.508	$^3B_1$	−55.2296	0	143.3	1.018

Source: Refs. 31–33.

[a]Method: SCF two-configuration calculation for $^1A_1$, single-configuration calculation for $^3B_1$. For silicon or nitrogen, a (9*s*6*p*3*d*) contracted Gaussian basis set; for hydrogen, a (4*s*2*p*) basis set. For experimental data on SiH_2, see Ref. 33.

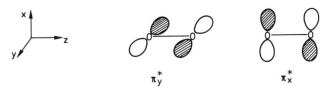

FIGURE 3-4.   Real molecular orbitals of O_2.

other and the Σ_g states are formed, with singlet or triplet multiplicity according to the respective spins. To summarize:

$$^3\Sigma_g^- \equiv |\pi_+^* \pi_-^*| \quad \text{(one spin component only)}$$

$$
\begin{array}{l}
\text{Imaginary}\\
\text{molecular}\\
\text{orbitals}
\end{array}
\quad
^1\Delta_g \equiv
\begin{cases}
|\pi_+^* \overline{\pi_+^*}| \\
|\pi_-^* \overline{\pi_-^*}|
\end{cases}
\qquad (3\text{-}18)
$$

$$^1\Sigma_g^+ \equiv |\pi_+^* \overline{\pi_-^*}| + |\pi_-^* \overline{\pi_+^*}|$$

It is instructive to enquire into the probability distribution of the electrons around the molecular axis.[37,39] In the ground triplet state the angular distribution of the two electrons is proportional to $\sin^2(\phi_1 - \phi_2)$, where ϕ is the azimuthal angle around the axis. The probability has a maximum for a 90° angular difference, when the electrons are in *orthogonal* planes.[37,39] Instantaneous repulsion between the electrons is minimized. In the highest singlet the distribution is $\cos^2(\phi_1 - \phi_2)$, the electrons being mostly in the *same* plane, with maximal electron repulsion. In the $^1\Delta_g$ state the electron repulsion is halfway between these two extremes, with no angular correlation [probability proportional to the modulus of $e^{i(\phi_1 - \phi_2)}$].

The picture with imaginary orbitals suffers precisely from the fact that such orbitals cannot be drawn out in space. By reverting to the real orbitals (3-16), it is simple to rephrase[37,40–42] the states by using (3-18) (for $^1\Delta_g$, one takes the sum and the difference divided by $2i$ of the two functions):

$$^3\Sigma_g^- \equiv |\pi_x^* \pi_y^*|$$

$$
\text{Real molecular orbitals} \quad
^1\Delta_g \equiv
\begin{cases}
|\pi_x^* \overline{\pi_x^*}| - |\pi_y^* \overline{\pi_y^*}| \\
|\pi_x^* \overline{\pi_y^*}| + |\pi_y^* \overline{\pi_x^*}|
\end{cases}
\qquad (3\text{-}19)
$$

$$^1\Sigma_g^+ \equiv |\pi_x^* \overline{\pi_x^*}| + |\pi_y^* \overline{\pi_y^*}|$$

These states are also drawn out in Table 3-3.[41–44] Equations (3-19) clearly reveal the diradical nature of ground triplet state and of one $^1\Delta_g$ state, whereas the other $^1\Delta_g$ state and the $^1\Sigma_g^+$ state have a "vertical-horizontal zwitterionic" character—

TABLE 3-3

Calculated Properties of O_2 (Experiment in Parentheses)

	Energy		Bond Length (Å)	Resonance Structures[b]	Alternative Description[c]
	Absolute (a.u.)	Relative (eV)			
$^1\Sigma_g^+$	−149.7383	1.79 (1.636)	1.258 (1.227)		
$^1\Delta_g$	−149.7672	1.031 (0.982)	1.242 (1.216)		
$^3\Sigma_g^-$	−149.8041	0	1.234 (1.207)		

Source: Refs. 41–44.

[a] Method: CI treatment with the use of "iterative natural orbitals" (Bender-Davidson scheme) (see Ref. 43) built on a double–ζ basis set including off-center *s* and *p* Gaussian functions. Experimental values from Ref. 44.
[b] See Ref. 41.
[c] See Ref. 42.

albeit of a local hole-pair type similar to that in CH_2. It should be noted that the two $^1\Delta_g$ components can freely exchange their diradical and zwitterionic character, as can be verified by rotating the x and y axes by $45°$ around the z axis (see the discussion of Table 3-11 in Section 3.9). The fact that, as mentioned earlier, the electrons are always in the same plane in the $^1\Sigma_g$ state and in orthogonal planes in the $^3\Sigma_g$ state is also borne out by these functions (Table 3-3). An alternative visual description of the $^1\Delta_g$ components is given by Goddard.[42]

The previous description of O_2 is useful for describing the reactions of O_2 with a reagent in the xz or yz plane and similar phenomena. But is there an alternative description, with the use of *atomic* orbitals π^A and π^B, valid for a reagent coming along the OO axis? Is there a zwitterionic state of the type[45]

$$\overset{\oplus}{O}\!\!-\!\!\overset{\ominus}{O} \leftrightarrow \overset{\ominus}{O}\!\!-\!\!\overset{\oplus}{O} \tag{3-20}$$

The answer can be obtained by simply expanding (3-19) together with (3-16). The results are

$$^3\Sigma_g^- = |A_xA_y| + |B_xB_y| - (|A_xB_y| + |B_xA_y|) \quad \text{(one triplet component)}$$

Atomic orbitals
$$^1\Delta_g = \begin{cases} (|A_x\bar{A}_x| - |A_y\bar{A}_y|) + (|B_x\bar{B}_x| - |B_y\bar{B}_y|) \\ \qquad + (|A_y\bar{B}_y| + |B_y\bar{A}_y|) + (|A_x\bar{B}_x| + |B_x\bar{A}_x|) \\ (|A_x\bar{A}_y| + |A_y\bar{A}_x|) + (|B_x\bar{B}_y| + |B_y\bar{B}_x|) \\ \qquad - (|A_x\bar{B}_y| + |B_y\bar{A}_x|) - (|A_y\bar{B}_x| + |B_x\bar{A}_y|) \end{cases}$$

$$^1\Sigma_g^+ = (|A_x\bar{A}_x| + |A_y\bar{A}_y|) + (|B_x\bar{B}_x| + |B_y\bar{B}_y|) \\ - (|A_y\bar{B}_y| + |B_y\bar{A}_y|) - (|A_x\bar{B}_x| + |B_x\bar{A}_x|)$$

$$\tag{3-21}$$

where the simplified notation $A = \pi_A$, $B = \pi_B$ has been used. All three singlet states *have an equal mixture of atom-atom covalent and atomic ionic character;* hence (3-20) is incorrect, describing at best 50% of any one state. In fact, a configuration interaction calculation shows that the "left-right" ionic character does not exceed 20%.* The explanation lies in the proper valence-bond description of O_2, which actually requires six electrons.[38] Covalent structures are required because they alone can lead to the proper dissociation products, $O(^3P) + O(^3P)$ for all four states as the bond distance increases (see also Fig. 2-1a); for instance

---

*The author is grateful to Dr. R. Weber for bringing this problem to his attention and to Dr. P. Hiberty for communicating unpublished calculations.

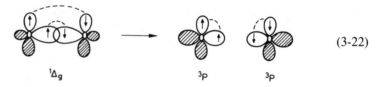

$$(3-22)$$

$${}^1\Delta_g \qquad {}^3p \qquad {}^3p$$

The scheme shown in (3-22) is impossible to draw out for ionic structures.

3.4   Twisted Ethylene or "Dimethylene" $\dot{C}H_2$—$\dot{C}H_2$ and Analogs

Twisted to 90°, ethylene forms a diradical that is clearly isoelectronic with O_2. In 1932 Mulliken, in his paper entitled "Quantum Theory of the Double Bond"[46] in which he interpreted the cis-trans isomerization of olefins, demonstrated the existence of four electronic states for twisted ethylene.

The orbitals available for the two odd electrons are p orbitals on the carbon centers:

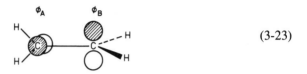

$$(3-23)$$

In terms of these atomic orbitals, called A and B for simplicity, the four states can be written (spatial part):

$$\begin{aligned}
{}^3\text{Diradical} &\equiv A(1)B(2) - B(1)A(2) &\quad {}^3A_2 \\
{}^1\text{Diradical} &\equiv A(1)B(2) + B(1)A(2) &\quad {}^1B_1 \\
{}^1\text{Zwitterion} &\equiv A(1)A(2) - B(1)B(2) &\quad {}^1B_2 \\
{}^1\text{Zwitterion} &\equiv A(1)A(2) + B(1)B(2) &\quad {}^1A_1
\end{aligned}$$

$$(3-24)$$

[compare with (1-2), (1-4), and (1-10)]. The symmetries in the D_{2d} point group are shown alongside. If delocalized molecular orbitals

$$\psi_+ = A + B$$
$$\psi_- = A - B$$

$$(3-25)$$

are constructed, the proper configurations then have the form shown in (3-8), and the detailed functions are shown further in Table 3-5.

TABLE 3-4

Calculated Properties of Dimethylene ($\dot{C}H_2 - \dot{C}H_2$)[a]

Symmetry (D_{2d})	Energy Absolute (a.u.)	Relative (eV)	Bond Length (Å)	Resonance Structures
$^1B_2$	−78.0060	3.51	1.43	
$^1A_1$	−78.0089	3.43	1.41	(+ covalent terms)
$^3A_2$	−78.1320	0.08	1.48	
$^1B_1$	−78.1349	0[b]	1.48	

Source: Ref. 47.

[a]Method: all-valence double-excitation configuration interaction calculation with the use of a double–ζ basis set including diffuse functions and polarization functions.

[b]Energy value 2.71 eV relative to ground-state ethylene.

The energies and ordering of these states[47,48] are given in Table 3-4. In the simple two-electron two-orbital picture, the triplet diradical state (two-electron energy $J_{AB} - K_{AB}$) should come below the singlet diradical state (energy $J_{AB} + K_{AB}$), as required by Hund's rule (Section 7.2).[49] The fact that a large configuration interaction calculation[47,48] reverses the order requires explanation[50] (see Chapter 7). Another surprising reversal is that between out-of-phase and in-phase zwitterionic states, with respective energies $J_{AA} - K_{AB}$ and $J_{AA} + K_{AB}$ in the two-electron model. Configuration interaction stabilizes the second state more than the first. Because of its symmetric nature, the ionic $^1A_1$ function can mix in covalent components[51] of the type

$$a(1)b(2) + b(1)a(2) \qquad (3\text{-}26)$$

where a and b are, for instance, $3s,4s$ or Rydberg s orbitals. Hence the in-phase zwitterionic state develops a small amount of covalent character, a possibility forbidden to the out-of-phase state because of its odd character relative to the C_2 axis perpendicular to the CC bond.

Another important problem concerns the correlation between the states of the diradical and those of the parent ethylene molecule.[52–54] The situation is quite confused because of the presence of the vertical excited Rydberg states. Essentially the diradical singlet state correlates with ground state ethylene, whereas the triplet diradical state originates from the valence triplet $^3\pi,\pi^*$ state of parent molecule. Finally, the $^1\pi,\pi^*$ state correlates with $(-)$ zwitterionic state while the doubly-excited $^1(\pi,\pi^*)^2$ state is related to the other $(+)$ zwitterionic state. These correlations are illustrated in Table 3-5. *Mixing* of the two molecular orbital determinants (both of which are 50% covalent and 50% ionic) is required to obtain the pure singlet diradical and $(+)$ zwitterionic states. Figure 3-5 gives a qualitative picture of the potential energy curves when the Rydberg states are included (see also Section 5.8).

By comparison with twisted ethylene, twisted stilbene (Table 3-6) shows[55] a large singlet-triplet gap for the diradical states. The relative stabilization of the singlet must be due to resonance of the type

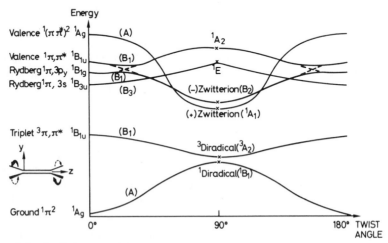

$$(3-27)$$

which appears because the phenyl substituents destroy the orthogonal symmetry planes existing in twisted ethylene, and which is forbidden to the triplet diradical.

FIGURE 3-5. Qualitative potential-energy diagram for ethylene as a function of twist angle (triplet Rydberg states are not shown). The dotted lines indicate an avoided crossing (see Section 5.8). Point groups are respectively D_{2h} (0°), D_2 (intermediate angles), and D_{2d} (90°).

TABLE 3-5

Correlation Between Planar Ethylene and Dimethylene[a]

Parent Molecule		→	Diradical	
State[b]	Function		Function [See (3-8)]	State
Ground (N)	$\lvert\pi\bar{\pi}\rvert$		$\lvert\psi_+\bar{\psi}_+\rvert - \lvert\psi_-\bar{\psi}_-\rvert$	Singlet diradical
Lowest triplet (T)	$\lvert\pi\pi^*\rvert$		$\lvert\psi_+\psi_-\rvert$	Triplet diradical
Lowest-valence singlet (V)	$\lvert\pi\bar{\pi^*}\rvert + \lvert\pi^*\bar{\pi}\rvert$		$\lvert\psi_+\bar{\psi}_-\rvert + \lvert\psi_-\bar{\psi}_+\rvert$	(−) Zwitterionic
Doubly excited singlet (Z)	$\lvert\pi^*\bar{\pi^*}\rvert$		$\lvert\psi_+\bar{\psi}_+\rvert + \lvert\psi_-\bar{\psi}_-\rvert$	(+) Zwitterionic

[a]Compare with the results presented in Table 5-1.
[b]See Ref. 53.

TABLE 3-6

Calculated Properties of Twisted Stilbene (1,2-Diphenyl Dimethylene)[a]

Symmetry $(C_2)^b$	Energy		Resonance Structures
	Absolute (a.u.)	Relative (eV)	
^1B	−530.4115	5.59	
^1A	−530.4338	4.98	(+ covalent terms)
^3B	−530.5773	1.07	
^1A	−530.6168	0[c]	(+ ionic terms)

Source: Ref. 55.

[a]Method: SCF calculation followed by configuration interaction; the orbital basis is a minimal STO-3G basis and the CI includes all important excitations within the subspace of the π orbitals of the molecule.
[b]Relative to twisted ethylene the only remaining symmetry element is a C_2 axis perpendicular to the CC bond; A_1 and B_1 states previously correlate with A here, and A_2 and B_2 correlate with B.
[c]Energy value 2.99 eV relative to ground-state stilbene.

3.5   Ozone

The valence-bond description of ozone[56] is illuminating. We start with the available *p* orbitals on the oxygen atoms, all pointing perpendicular to the OO bonds. The lower *sp* hybrids on the terminal oxygen atoms and the sp^2 orbital on the central atom, each filled with an electron pair, are omitted. The lowest-energy construct then has 4 $p\pi$ electrons and 4 $p\sigma$ electrons:

$$(3\text{-}28)$$

In molecular orbital language the four π electrons would occupy two π orbitals, the higher of which would have nonbonding "allylic" character and be localized on the terminal oxygens, with one electron on each oxygen (see Section 3.8). Regardless of the description used, the ozone molecule thus appears as a diradical with two odd π electrons: a "π,π" diradical. However, contrary to the cases discussed in Sections 3.2–3.4, the two odd orbitals have the same local symmetry and can overlap, either "directly" or "indirectly" through the central atom.* In resonance language, this is characterized by the admixture of doubly–bonded structures:

$$(3\text{-}29)$$

(π electrons are represented by dots above the atoms and σ electrons, by dots below sideways). The singlet diradical state lies lower in energy than the triplet diradical state, for which such resonance is forbidden. The resonance structures on the right correspond to the major components of a high-lying zwitterionic singlet state (third A_1 state; see Table 3-7).

Note that in the horizontal σ plane the two p-σ lone pairs point at each other, leading to a four-electron repulsion similar to that between two helium atoms. This repulsion could be relieved by exciting a σ electron into the π system. The molecule remains a diradical but is now an excited diradical:

---

*In the GVB picture (Section 1.4) the two singly–occupied terminal π orbitals readjust to become orthogonal to the central π orbital. To do so, they borrow some character from precisely this orbital, and their overlap increases as a consequence.

$$(3\text{-}30)$$

Indeed, the σ four-electron repulsion has now been replaced by an even costlier π four-electron repulsion between neighboring oxygens. Nevertheless, this "σ,π" diradical possesses a group of four diradical and zwitterionic states described in the usual manner (Section 3.1). Furthermore, the diradical states are split into two because of the two identical structures of type (3-30) which can be written. The ionic states ($\sigma^{\oplus},\pi^{\ominus}$ or $\sigma^{\ominus},\pi^{\oplus}$) are also split into two, but the latter pair, with 4 π electrons, are identical with the zwitterionic states built from the π,π diradical, whereas the first pair is identical with those built from the σ,σ diradical.

Hence ozone has dual diradical possibilities; to each diradical system, the π,π one and the σ,π one, corresponds a series of states (see Table 3-7). The existence of a high-energy σ,σ diradical, with six π electrons (two σ electrons have been promoted to the π system) should also be mentioned. This diradical, in its singlet state, forms *triangular* ozone[58] (see also Fig. 2-1c). The corresponding covalent resonance structure

$$(3\text{-}31)$$

is the major component of the lowest excited 2^1A_1 state. There is also a slightly lower σ,σ diradical triplet state with the two unpaired spins in the molecular plane.

3.6 Trimethylene $\dot{C}H_2$—CH_2—$\dot{C}H_2$ and Analogs

The trimethylene diradical, postulated in 1958 to occur as an intermediate in the geometrical isomerization of cyclopropane,[59] is isoelectronic with ozone. But replacement of an oxygen atom by a CH_2 group automatically mobilizes a p orbital that was free on the oxygen atom. Hence the number of available p orbitals is reduced to 2, one on each terminal CH_2 group:

$$(3\text{-}32)$$

TABLE 3-7
Calculated Properties of Ozone[a]

Nature of Odd Orbitals and of State	Symmetry (C_{2v})	Energy Absolute (a.u.)	Energy Relative (eV)	Geometry R = OO bond (opt. for $\theta = 116.8°$) (Å)	Geometry $\theta = \widehat{OOO}$ angle (opt. for $R = 1.376$ Å) (deg.)	Resonance Structures Major	$\leftrightarrow$	Resonance Structures Minor
σ,σ and $\sigma^{\oplus},\pi^{\ominus}$ (Zw.) $\;\;2^1B_2$	2^1B_2	—	—	—	—			
4^1A_1	4^1A_1	—	—	—	—			
π,π and $\sigma^{\ominus},\pi^{\oplus}$ (Zw.)	3^1A_1	−224.3119	5.52	1.50	—			
π,π and $\sigma^{\ominus},\pi^{\oplus}$ (Zw.)	$^1B_2$	−224.3546	4.36	1.48	117			
σ,σ (Dir.)	2^1A_1	−224.4255	2.43	1.49	60			

		State	Energy				Structures
σ,σ	(Dir.)	2^3B_2	-224.4350	2.17	1.49	107	
σ,π	(Dir.)	$\left\{\begin{array}{l}{}^1B_1\\{}^1A_2\end{array}\right.$	-224.4674 / -224.4696	1.29 / 1.23	1.44 / 1.44	112 / 101	
σ,π	(Dir.)	$\left\{\begin{array}{l}{}^3A_2\\{}^3B_1\end{array}\right.$	-224.4744 / -224.4762	1.10 / 1.05	1.45 / 1.44	100 / 114	
π,π	(Dir.)	$^3B_2$	-224.4865	0.77	1.44	110	
π,π	(Dir.)	$^1A_1$	-224.5148	0	1.38	112	

Source: Ref. 57.
Method: generalized valence-bond, double–ζ basis, extensive CI (π electrons are represented by dots above the atoms and σ electrons, by dots below sideways).

Clearly, π,π, σ,π, or σ,σ diradical forms are still possible; but instead of being different states of a single molecular geometry, they now correspond to different *conformers* that may be on the potential surface of a single state:

π,π	π,σ	σ,π	σ,σ
(Edge to edge)	(Edge to face)	(Face to edge)	(Face to face)

(3-33)

Immediately we expect the triplet diradical state to be the ground state in the σ,π conformers with orthogonal odd orbitals as in methylene.* In the π,π and σ,σ conformers weak direct overlap and through-bond conjugation[60] can, in principle, stabilize the singlet:

(3-34)

For a given $\widehat{CCC}$ angle (113°, close to the values that optimize the π,π and σ,π diradical singlets), the σ,σ diradical singlet has the lower energy—an indication of its tendency to collapse to cyclopropane without any energy barrier (see Table 3-8).

A good deal of debate has surrounded the question of whether trimethylene is an intermediate, in a secondary minimum on the potential surface (see Refs. 5 and 6 in Chapter 2 and discussion of Fig. 2-2b in Section 2.1) in the geometric isomerization reaction of cyclopropane.[3,59,61,62] In the light of the most detailed experimental evidence available,[63] the issue is not resolved. However, apart from the possibility of small local conformational minima, theory gives no evidence for such a secondary minimum.

If the central CH_2 group of trimethylene is substituted by NH or oxygen, the ionic resonance structures gain weight in the π,π diradical:

---

*The special stabilization of singlet due to spin polarization[50] observed in dimethylenes (page 73) should, if anything, favor triplet state here because in trimethylene the polarization must carry through an extra bond.

TABLE 3-8
Calculated Energies for Trimethylene (Diradical States Only)[a]

"Conformation" and Odd Orbitals	Symmetry	Energy		Geometry	
		Absolute (a.u.)	Relative (eV)	$\widehat{CCC}$ (deg.)	CC (Å)
σ,π (C_s)	^1A	-116.7535	0.18	112.4	$1.538\ (C_{central}\text{—}C_\sigma)$ $1.513\ (C_{central}\text{—}C_\pi)$
	^3A	-116.7588	0.03	—	—
π,π (C_{2v})	$^1A_1$	-116.7544	0.16	113.9	1.522
	$^3B_2$	-116.7575	0.07	—	—
σ,σ (C_{2v})	$^3B_2$	-116.7588	0.03	—	—
	$^1A_1$	-116.7601	0^b	$[113]^c$	1.522

Source: Ref. 61.
[a]Method: open-shell restricted SCFMO, single–ζ basis, 3×3 CI.
[b]Energy value 2.1 eV above cyclopropane.
[c]Fixed.

$$(3\text{-}35)$$

and a singlet ground state establishes itself unambiguously.[64] In the σ,π conformation, however, such resonance is reduced to one ionic structure, that in which the methylene group carrying the σ orbital is the anionic center. In the σ,σ conformation such resonance disappears entirely and we are left with the direct throughspace hole-charge resonance structures in (3-34).

Hence the singlet-triplet separation is expected to be smallest in the σ,σ ("face-to-face") conformations, largest in the π,π ("edge-to-edge") conformations and intermediate in the σ,π ("edge-to-face") conformations.

Finally, if the terminal CH_2 groups are substituted by oxygen atoms, we recover the multiple states built on p orbitals as in ozone. The lack of a central oxygen π orbital relay, however, renders the σ,σ and σ,π diradicals more competitive in such dioxyl diradicals.[65]

3.7 Vinylmethylene $\overset{..}{C}H$—CH=CH$_2$

Vinylmethylene, postulated as an intermediate in the ring opening of cyclopropene[66] and that can be obtained by decomposing or irradiating vinyl diazo compounds,* is somewhat intermediate between ozone and trimethylene. The terminal CH center has both a σ odd orbital and a π odd orbital available, like an oxygen atom, whereas the terminal methylene CH$_2$ group has a single p orbital available. The possibility of conjugation through the central CH group offers a novel feature.

Depending on the conformation of the terminal methylene group, the system can be a planar σ,π diradical

$$(3\text{-}36)$$

or an "orthogonal" σ,σ diradical that is essentially an open cyclopropene

$$(3\text{-}37)$$

Interestingly, in the σ,π diradical the two resonance forms are not equivalent and do not have equal weights.[69] In the triplet diradical state the first form is favored; in the singlet diradical state the second form is preferred.

$$(3\text{-}38)$$

Triplet   Singlet

These preferences can be traced back to the two-electron energies of singlet and triplet [see (3-8)]:

$$E_{\text{singlet}} = J_{ab} + K_{ab}$$

$$E_{\text{triplet}} = J_{ab} - K_{ab}$$

$$(3\text{-}39)$$

*See Ref. 67 and the article by Chapman et al. quoted therein (Footnote 11a); see also Ref. 68.

The Coulomb integral that pushes the electrons apart tends to favor a 1,3 diradical, but its variation with distance is relatively slow, and as the π electron approaches atom 1, the repulsion with the σ electron there must be partially offset by the specially strong nuclear attraction as a result of σ unsaturation at that carbon atom. Hence the exchange integral apparently dominates the electronic behavior. In the usual manner the triplet maximizes its (negative) exchange energy by placing the two electrons on the same atom (see also Chapter 7). The diradical singlet state minimizes its positive exchange contribution (K_{13}) but does not fare too well and is higher in energy than is the "zwitterionic" singlet, in which both odd electrons are paired in the low-lying σ orbital (lowest singlet $1^1A'$ of the planar form).

Not surprisingly, the σ,σ diradical (with the $\widehat{CCC}$ angle fixed at 120°, to avoid collapse to cyclopropene) has nearly degenerate singlet and triplet ground states because of the ionic admixture that helps the singlet. The energies of both con-formers[69] are shown in Table 3-9. The ring opening of cyclopropene should lead to the singlet diradical state ($1^1A'$), which can stabilize itself by methylene rotation. If the electron distribution were to remain unchanged, this twisting motion would lead to the $^1A''$ diradical state of the planar form. In practice, the electrons *rearrange* and the molecule ends up in the planar $^1A'$ state, with local ion-pair character (further decay to triplet is possible). Such a rearrangement implies an "avoided crossing" (see Section 5.8) between the potential energy surfaces for "diradical" distribution of the electrons ($1^1A'$, orthogonal $\rightarrow$ $^1A''$, planar) and for "zwitter-ionic" distribution of the electrons ($2^1A'$, orthogonal $\rightarrow$ $1^1A'$, planar).

3.8 The Allyl Radical $\dot{C}H_2—CH=CH_2$

We turn for a change to a monoradical system. Although it possesses only one odd electron, the allyl radical is important because its states illustrate nicely the differ-ence between in-phase and out-of-phase resonance (Section 1.3). The description of these states is given by Levin and Goddard.[70]

Conventionally, resonance in allyl occurs between the two structures in which the odd electron occupies one or the other terminal CH_2 centers:

$$\text{//}\diagdown \leftrightarrow \diagup\text{\textbackslash\textbackslash} \tag{3-40}$$

The corresponding wave functions can be combined positively or negatively. The *lower energy combination is the negative one*

$$\text{//}\diagdown \overset{(-)}{\leftrightarrow} \diagup\text{\textbackslash\textbackslash} \equiv \diagup\diagdown \tag{3-41}$$

which can be shown[70] to be identical to a single coupling scheme of the three odd electrons [see RHS of (3-41)], with "triplet coupling" between the termini. The

TABLE 3-9
Calculated Energies for Vinylmethylene ($\ddot{C}H—CH—CH_2$)[a]

State	Energy: Absolute (a.u.)	Energy: Relative (eV)	Major Resonance Structure	State	Energy: Absolute (a.u.)	Energy: Relative (eV)	Major Resonance Structure
$2^1A'$	—	—	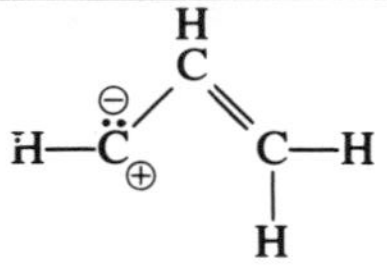	$3^1A'$	—	—	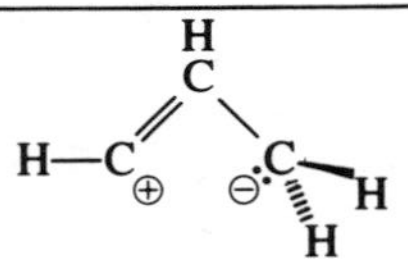
$^1A''$	−115.8591	0.69	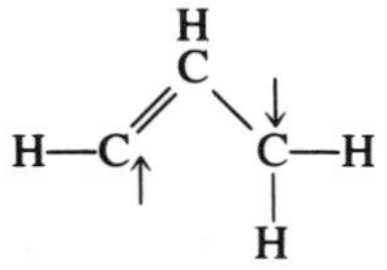	$2^1A'$	—	—	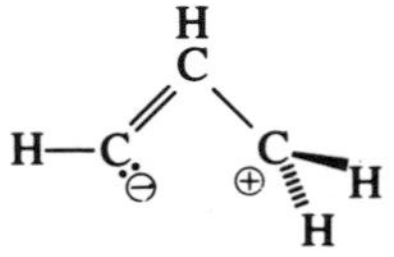
$1^1A'$	−115.8828	0.51	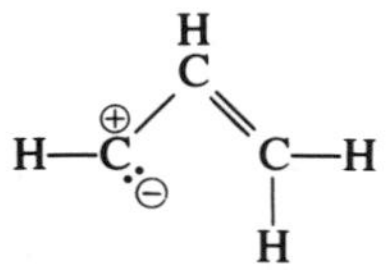	$1^1A'$	−115.8731	0.77	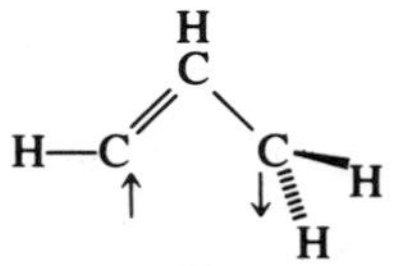
$^3A''$	−115.9015	0^b	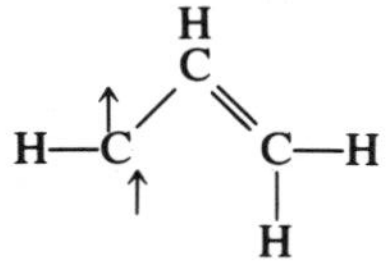	$^3A'$	−115.8736	0.76	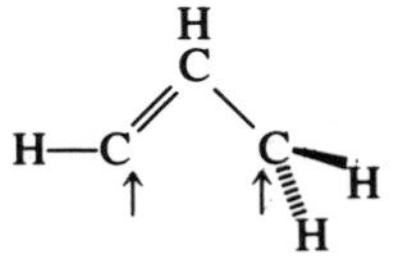
	Planar (C_s symmetry)				Orthogonal (C_s symmetry)		

Source: Ref. 69.
[a]Method: GVB, double–ζ basis, CI.
[b]Energy value 1.07 eV above cyclopropene.

higher-energy combination is equivalent to the alternative, linearly independent coupling scheme with "singlet coupling" between the termini (see also Table 1-1):*

$$\nearrow\!\!\!\!\searrow \cdot \overset{(+)}{\longleftrightarrow} \cdot \nearrow\!\!\!\!\searrow \;\equiv\; \uparrow\!\!\!-\!\!\!-\!\!\!\downarrow \qquad (3\text{-}42)$$

The absence of π bonding between adjacent carbons and the presence of a long bond explains the high energy.

Comparison with the molecular orbital description of allyl[71] shows (3-41) to correspond roughly to a configuration $\psi_1^2\,\psi_2$, where ψ_1 is the bonding π orbital and ψ_2, the non-bonding molecular orbital that puts odd spin on the terminal atoms. The correspondence is not exact, however, since $\psi_1^2\psi_2$ contains ionic terms such as

$$\nearrow\!\!\!\!\searrow \overset{\oplus}{}\underset{\ominus}{} \qquad (3\text{-}41a)$$

Structure (3-42), on the other hand, corresponds roughly to the excited configuration $\psi_1\,\psi_2^2$ in which two paired electrons are located on the terminal atoms, mixed in with $\psi_1^2\psi_3$ to eliminate the ionic terms.

The energies of the two lowest doublet states of allyl are shown in Table 3-10. From these numbers a resonance energy of approximately 47 kJ/mole is deduced, whereas an *antiresonance* energy of ~250 kJ/mole is obtained.[70] This antiresonance

TABLE 3-10
Calculated Energies for Allyl[a]

| Symmetry (C_{2v}) | Energy | | Resonance Structures |
	Absolute (a.u.)	Relative (eV)	
$^2B_1$	−116.3034	3.20	$\nearrow\!\!\!\!\searrow \cdot \overset{(+)}{\longleftrightarrow} \cdot \nearrow\!\!\!\!\searrow$
$^2A_2$	−116.4211	0	$\nearrow\!\!\!\!\searrow \cdot \overset{(-)}{\longleftrightarrow} \cdot \nearrow\!\!\!\!\searrow$

Source: Ref. 70.
[a]Fixed geometry, R_{CC} = 1.40 Å, R_{CH} = 1.08 Å, all angles = 120°.

*The reader will easily check, for instance, that, for A—B—C, $|(\phi_A\overline{\phi_B} + \phi_B\overline{\phi_A})\phi_C| + |\phi_A(\phi_B\overline{\phi_C} + \phi_C\overline{\phi_B})| = |(\phi_A\overline{\phi_C} + \phi_C\overline{\phi_A})\phi_B|$ through the cancellation of the first and fourth terms on the left-hand side.

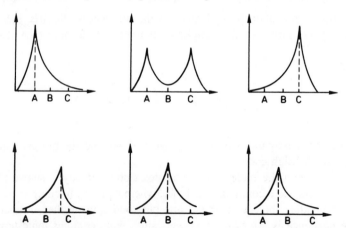

FIGURE 3-6.   Opposite behavior of unpaired spin density *(above)* and electron pair probability *(below)* in the bond shift process of allyl.

energy, defined as the destabilization energy due to in-phase resonance (3-42), is relatively large in the same way that antibonding orbitals are more antibonding than bonding orbitals are bonding. The next-higher excited states are probably Rydberg states ($2p\pi \rightarrow 3p\pi$, etc.).

Instead of considering resonance between structures with the same geometry, consider now the actual *bond shift process**:

$$\overset{\uparrow}{A}\diagdown\overset{B}{\diagup}\diagdown_C \longrightarrow A\diagdown\overset{B}{\diagup}\diagdown\overset{\uparrow}{C} \tag{3-43}$$

This reaction is a simple model for an electron transfer process (see Section 8.9). In this process the unpaired π spin density transfers from left to right, whereas the π pair probability—determined by the single coupling scheme (3-41)—simply shifts in the other direction.[74] The behaviors are depicted schematically in Fig. 3-6. It should be noted that, since the electrons are indistinguishable, it is impossible to ascertain whether the final electron on carbon is the same as the initial electron on A or the electron of appropriate spin initially in bond BC.

3.9   Cyclobutadiene†

Cyclobutadiene, which eluded synthesis until 1972,[76] is of particular interest because it is a diradical of high symmetry. A reminder of the π-orbital energy pattern and orbital symmetries of the molecule is given in Fig. 3-7. The two odd electrons

---

*The potential energy surface for such a process, which has a minimum for symmetric or very nearly symmetric allyl, is illustrated in Fig. 2 of Paldus and Veillard,[72] who also study the "instability" of the Hartree-Fock solutions. See also Ref. 73.

†For a review of the molecules pertinent to this and the following section, see Ref. 75.

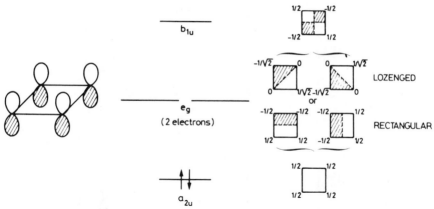

FIGURE 3-7. π-Orbital energies and orbital amplitudes in cyclobutadiene (D_{4h} symmetry).

occupy a pair of nonbonding molecular orbitals. Because of their orbitally degenerate (e_g) symmetry, these orbitals can be described in a number of ways:

A pair of *"rectangular"* orbitals with nodal planes cutting through pairs of opposite CC bonds. (These are actually the appropriate orbitals to consider for distortion of the molecule from square-planar to rectangular.)

A pair of *"lozenged"* orbitals with nodal planes cutting through opposite carbon atoms. (These are appropriate to describe distortion from square to planar kite.)

Table 3-11 shows that the distribution of the odd electrons in the two orbitals, for the low-lying states of the molecule, depends on whether the rectangular description or the lozenged description is used.

If we adhere to the conventional view that a diradical has two odd electrons in two different orbitals, we see that the "first" singlet of cyclobutadiene is a "diradical" if the lozenged description is used but has "zwitterionic" character if the rectangular description is used. The situation is reversed for the "second" singlet. For the third singlet, the two formalisms agree. This situation bears very close analogy to the singlet states of oxygen [see (3-19) and its discussion]. To discover the true nature of the wave functions, it is better to write them out in terms of the atomic orbitals (Table 3-11). The first singlet is then purely covalent; the second and third singlet are mainly ionic but contain small diagonal (atom 1 to atom 3) covalent terms.

If the wave functions of cyclobutadiene resemble those of O_2, the energetics, however, resemble more those of dimethylene (Section 3.4). The two-electron energies[75] are shown in Table 3-11. The rectangular description is characterized by very close Coulomb integrals J_{rr} and J_{rs} (identical in the zero-differential overlap approximation; see Section 1.9) and a large exchange integral K_{rs}, whereas the lozenged description has J_{ll} larger than J_{lm}, and a very small exchange integral K_{lm} (zero for zero-differential overlap). Hence the two-electron picture puts the triplet only slightly below the lowest singlet, as would want Hund's rule. A more so-

TABLE 3-11

Odd-Electron Distribution for the Four Diradical States of Cyclobutadiene[a]

State	Energy	Rectangular orbitals (r,s)	Lozenged orbitals (ℓ,m)	Wave function $(a,b,c,d\ \text{atomic orbitals})$
$(^1A_{1g})$ "Third" singlet	$J_{rr} + K_{rs}$ $= J_{\ell\ell} + K_{\ell m}$			$\|a\bar{a} + c\bar{c}) + (b\bar{b} + d\bar{d}\|$ $- \|a\bar{c} + c\bar{a}) + (b\bar{d} + d\bar{b}\|$
$(^1B_{2g})$ "Second" singlet	$J_{rs} + K_{rs}$ $= J_{\ell\ell} - K_{\ell m}$			$\|a\bar{a} + c\bar{c}) - (b\bar{b} + d\bar{d}\|$ $- \|a\bar{c} + c\bar{a}) - (b\bar{d} + d\bar{b}\|$
$(^1B_{1g})$ "First" singlet	$J_{rr} - K_{rs}$ $= J_{\ell m} + K_{\ell m}$			$\|a\bar{b} + b\bar{a}) + (c\bar{d} + d\bar{c}\|$ $- \|a\bar{d} + d\bar{a}) + (b\bar{c} + c\bar{b}\|$
$(^3A_{2g})$ Triplet diradical	$J_{rs} - K_{rs}$ $= J_{\ell m} - K_{\ell m}$			$\|ab + cd\| - \|ad + bc\|$

[a] See Section 5.4 for a more detailed description of the $^1A_{1g}$ state.

TABLE 3-12
Calculated Energies for Cyclobutadiene[a]

State (D_{4h})	Energy		CC Bond Length (Å)
	Absolute (a.u.)	Relative (eV)	
$^1B_{2g}$	−153.6195	3.21	—
$^1A_{1g}$	−153.6655	1.95	—
$^3A_{2g}$	−153.7197	0.47	1.428
$^1B_{1g}$	−153.7372	0[b]	1.423

Source: Ref. 80.

[a]Method: two-configuration SCF with the use of "6-31G" Gaussian basis, followed by CI including all relevant single and double excitations and corresponding to both σ- and π-electron correlation.
[b]Gains another 0.52 eV by distorting to rectangular form with CC bond lengths of 1.334 and 1.564 Å.

phisticated calculation reverses the order[77,78] and puts the singlet firmly below the triplet. Indeed, configuration interaction between the electronic configuration ^1lm for the lowest singlet and an excited configuration in which one bonding a_{2u} electron is excited to the antibonding b_{1u} orbital (Fig. 3-7) allows the singlet to "*spin polarize.*" At any given instant the α electron in the a_{2u} pair tends to be at the opposite corner from the α member of ^1lm, and the same holds for the two β electrons (see also Fig. 7-6). In this manner the Coulomb repulsion between the nonbonding electrons and the bonding pair is slightly reduced. The overall spin density is precisely that shown in Fig. 3-1.

The ground singlet is further stabilized by a "pseudo" or "second-order" Jahn-Teller nuclear distortion.[71] The close energy of all three singlets offers the possibility to the molecule of lowering its energy by distorting along a vibrational coordinate whose symmetry mixes ground and excited states. Here a b_{1g} vibration, which converts square cyclobutadiene to rectangular cyclobutadiene, effectively lowers the energy by mixing ground $^1B_{1g}$ and excited $^1A_{1g}$ states. The energy gain for ground singlet is roughly of the same order of magnitude as that due to the spin polarization (Section 7.3).

Another analogy between cyclobutadiene and dimethylene appears in the lowering of the $^1A_{1g}$ ($^1A_1$ in ethylene) "in-phase" zwitterionic state as a result of mixing with higher excited covalent configurations. This "third" singlet actually falls below the "second" singlet $^1B_{1g}$. The calculated energies[79,80] are shown in Table 3-12, and the detailed process is discussed in Section 5.4.

3.10 Trimethylenemethane[75]

Like cyclobutadiene, trimethylenemethane offers four trigonal carbon centers to accommodate two odd electrons. It also presents the difficulty of having different conformers:

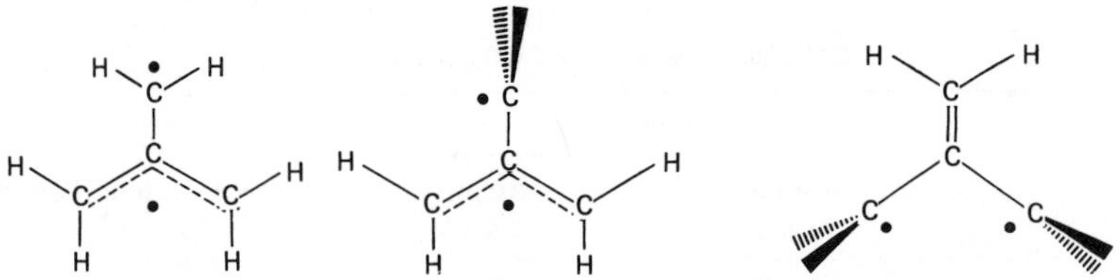

| | | (3-44) |

Coplanar
(π,π diradical)

Orthogonal
(σ,π diradical)

Bis-orthogonal
(open methylene-cyclopropane:
σ,σ diradical)

Interest has focussed on the coplanar conformer, particularly since evidence of racemization of chiral methylenecyclopropanes through some fully coplanar, achiral intermediate—probably a singlet since oxygen has no effect on the stereochemical characteristics of the rearrangement—has been obtained.[81] Again, the competition between triplet and singlet diradical states has been of primary concern.

The orthogonal diradical can be dealt with first since it is a perfect analog of dimethylene in which one terminal orbital has delocalized onto two methine substituents:

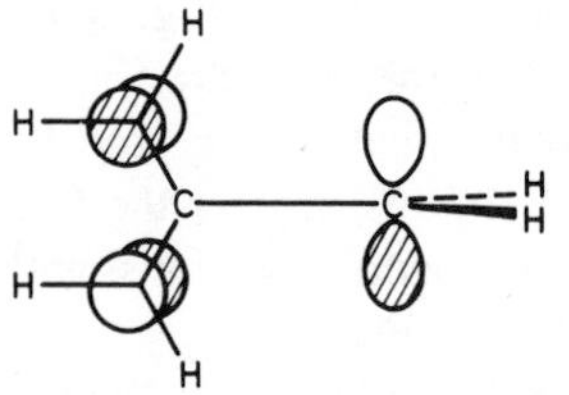

$$(3-45)$$

[compare with (3-23)]. Indeed, as in that system, the singlet and triplet diradical states lie close together.

In a first approximation, both singlet and triplet are expected to be stabilized by methylene rotation into the coplanar form, because of the resonance

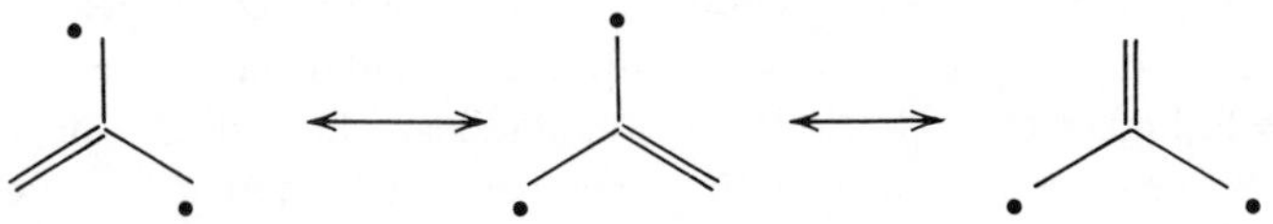

$$(3-46)$$

which extends onto one more structure than in the orthogonal case. If the prediction turns out to be correct for the triplet state, it is false for the singlet! The triplet descends, whereas the singlet rises (Fig. 3-8).[82] The reason can be traced back to the nature of the odd molecular orbitals, as shown in (3-47):

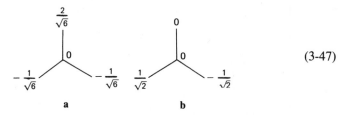

$$(3\text{-}47)$$

a   b

These odd orbitals form a degenerate pair that always—regardless of the combination chosen—have a common amplitude on certain terminal carbon atoms.[83] This common amplitude is an asset to the triplet state because of the large negative exchange integral appearing in its energy ($J_{ab} - K_{ab}$). But the singlet diradical state (energy $J_{ab} + K_{ab}$) is pushed up relative to the orthogonal form because K_{ab} will involve large self-repulsion energies when two electrons appear on the same atom.

Hence the localizability of the two nonbonding molecular orbitals on different sets of atoms (possible in cyclobutadiene, ground singlet, but impossible in trimethylenemethane, ground triplet) is a crucial factor in determining the ground state of conjugated diradicals.[84]

The planar singlet of trimethylenemethane, however, can relieve a good part of its large electronic repulsion energy, in two different ways. Borden[85] has been instrumental in showing how first an appropriate configuration interaction can push the nonbonding electrons into different parts of the molecule (a single *p* orbital, and an allylic orbital, just as if the molecule were in its orthogonal form) and also how a first-order Jahn-Teller distortion,[71] for instance, by lengthening one CC bond, can change a highly symmetric D_{3h} trimethylenemethane in its degenerate $^1E'$ state into a weakly coupled (allyl, methylene) system. Both effects, as shown first by Schaefer for the latter,[86] will relieve the large electron repulsion in the singlet. Early calculations (including this author's own work),[83] which neglected some of these effects, had placed the coplanar singlet much too high relative to the triplet. Its true position[87,88] now appears to be relatively close to the orthogonal singlet (Table 3-13). The singlet surface is thus very "floppy" since the orthogonal singlet diradical apparently needs no more than 10–14 kJ/mole to reclose to methylenecyclopropane or twist to planar diradical.[88]

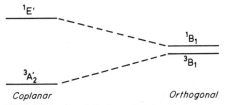

FIGURE 3-8.   Opposite qualitative behavior of singlet and triplet diradical states of trimethylenemethane under twist of a methylene group.[82] Note that the planar singlet will find some stabilization from a distortion to C_{2v} symmetry.

TABLE 3-13
Calculated Energies for Trimethylenemethane $C(CH_2)_3$ [a]

Coplanar geometry diagram (atoms labeled 1, 2, 3, 4)

Orthogonal geometry diagram: $118.8°$, 1.074, 1.500, $117.4°$, 1.589, $121.2°$, 1.074, 1.074

	State (D_{3h} symmetry)	Energy Absolute (a.u.)	Energy Relative (eV)	Geometry (Å)	State (C_{2v} symmetry)	Energy Absolute (a.u.)	Energy Relative (eV)	Geometry
Zwitterionic	$^1A_1$	—	—		$^1A_1$	—	—	
	$^1A_1$	—	[5.43]		$^1A_1$	—	[5.94]	
Diradical	$^1E'(D_{3h})$ distorts to	—	[1.14]	$C_1C_4 = 1.446$				
	$-{}^1B_2(C_{2v})$	—	[0.96]	$C_3C_4 = C_2C_4 = 1.402$; $C_1C_4 = 1.540$				
	or							
	$-{}^1A_1(C_{2v})$	—	[0.92]	$C_1C_4 = C_2C_4 = 1.497$; $C_3C_4 = 1.355$	$^1B_1$	-154.9643	$0.62 \rightarrow$	
	$^3A'_2$	-154.9871	0	$C_1C_4 = 1.429$	$^3B_1$	—	[0.53]	
	Coplanar				Orthogonal			

Source: Refs. 82, 85, 87.

[a] Method (absolute energies): SCF with double–ζ followed by CI including all single and double excitations. The absolute energies for planar triplet and orthogonal singlet diradicals are obtained from Ref. 82, using $^3A'_2$ and $^1B_1$ as reference states, whereas the coplanar geometries and distorted diradical singlet energies are obtained from Davidson and Borden.[85] Ref. 88 gives an even smaller (10 kJ/mole) energy difference between planar $^1A_1$ singlet and orthogonal $^1B_1$ singlet than the Table (29 kJ/mole).

Despite this understanding of the singlet behavior, a frustrating problem remains: the discrepancy between othogonal singlet-planar triplet difference, as calculated (59.6 kJ/mole)[87] versus observed [upper limit of 14.6 kJ/mole[89] for species of type (3-48)].

$$\text{(3-48)}$$

References

1. J. A. Berson, *Acc. Chem. Res.* **11**, 466 (1978).
2. G. Kothe, K.-H. Denkel, and W. Summermann, *Angew. Chem. Internatl. Ed.* **9**, 906 (1970); A. Rassat and H. U. Sieveking, *Angew. Chem. Internatl. Ed.* **11**, 304 (1972).
3. R. Huisgen, *Angew. Chem. Internatl. Ed.* **2**, 565 (1963).
4. L. Salem and C. Rowland, *Angew. Chem. Internatl. Ed.* **11**, 92 (1972).
5. J. Michl, *J. Molec. Phot.* **4**, 257 (1972).
6. K. N. Houk in *Pericyclic Reactions*, Vol. 2, edited by A. P. Marchand and R. E. Lehr (Academic, New York, 1977) p. 181 (Figure 8).
7. P. Caramella and K. N. Houk, *J. Am. Chem. Soc.* **98**, 6397 (1976).
8. J. Cizek and J. Paldus, *J. Chem. Phys.* **47**, 3976 (1967); J. Koutecky, *J. Chem. Phys.* **46**, 2443 (1967); A. R. Gregory in *Chemical and Biochemical Reactivity, Proceedings Sixth Jerusalem Symposium*, edited by E. D. Bergmann and B. Pullman (Israel Academy of Sciences, Jerusalem, 1974), p. 23.
9. P. O. Löwdin, *Rev. Mod. Phys.* **35**, 496 (1963).
10. H. Fukutome, *Progr. Theor. Phys.* **47**, 1156 (1972); ibid. **49**, 22 (1973).
11. K. Yamaguchi, T. Fueno, and H. Fukutome, *Chem. Phys. Lett.* **22**, 461 (1973).
12. V. Bonacic-Koutecky and J. Koutecky, *Theoret. Chim. Acta. (Berl.)* **36**, 149 (1975); *Chemical and Biochemical Reactivity, Proceedings Sixth Jerusalem Symposium*, edited by E. D. Bergmann and B. Pullman (Israel Academy of Sciences, Jerusalem, 1974), p. 131.
13. K. Yamaguchi, *Chem. Phys. Lett.* **35**, 230 (1975).
14. K. Yamaguchi, *Chem. Phys. Lett.* **30**, 434 (1975).
15. D. Dohnert and J. Koutecky, *J. Am. Chem. Soc.* **102**, 1789 (1980).
16. C. F. Bender and H. F. Schaefer, *J. Am. Chem. Soc.* **92**, 4984 (1970); S. V. O'Neil, H. F. Schaefer, and C. F. Bender, *J. Chem. Phys.* **55**, 162 (1971).
17. P. C. H. Jordan and H. C. Longuet-Higgins, *Molec. Phys.* **5**, 121 (1962).
18. G. Herzberg and J. W. C. Johns, *Proc. Roy. Soc.* **A295**, 107 (1966).
19. R. R. Lucchese and H. F. Schaefer, *J. Am. Chem. Soc.* **99**, 6765 (1977).
20. B. O. Roos and P. M. Siegbahn, *J. Am. Chem. Soc.* **99**, 7716 (1977).
21. L. B. Harding and W. A. Goddard, *J. Chem. Phys.* **67**, 1777 (1977).
22. C. W. Bauschlicher and I. Shavitt, *J. Am. Chem. Soc.* **100**, 737 (1978).
23. C. W. Bauschlicher, H. F. Schaefer, and P. S. Bagus, *J. Am. Chem. Soc.* **99**, 7106 (1977).
24. J. F. Harrison, R. C. Liedtke, and J. F. Liebman, *J. Am. Chem. Soc.* **101**, 7162 (1979).
25. N. C. Baird and K. F. Taylor, *J. Am. Chem. Soc.* **100**, 1333 (1978).
26. D. Feller, W. T. Borden, and E. R. Davidson, *Chem. Phys. Lett.* **71**, 22 (1980).
27. R. L. Tyner, W. M. Jones, Y. Ohrn, and J. R. Sabin, *J. Am. Chem. Soc.* **96**, 3765 (1974).
28. H. Dürr and G. Scheppers, *Chem. Ber.* **100**, 3236 (1967).
29. H. Dürr, *Topics Curr. Chem.* **40**, 103 (1973).
30. R. Shepard, A. Banerjee, and J. Simons, *J. Am. Chem. Soc.* **101**, 6174 (1979).

31.   J. H. Meadows and H. F. Schaefer, *J. Am. Chem. Soc.* **98**, 4383 (1976).

32.   C. F. Bender, J. H. Meadows, and H. F. Schaefer, *Faraday Disc. Chem. Soc.* **62**, 59 (1977).

33.   I. Dubois, G. Herzberg, and R. D. Verma, *J. Chem. Phys.* **47**, 4262 (1967).

34.   J. E. Lennard-Jones, *Transact. Faraday Soc.* **25**, 668 (1929).

35.   G. N. Lewis, *J. Am. Chem. Soc.* **38**, 762 (1916); *Chem. Rev.* **1**, 231 (1924).

36.   R. S. Mulliken, *Phys. Rev.* **32**, 880 (1928); R. S. Mulliken, *Rev. Mod. Phys.* **4**, 1 (esp. pp. 54–56 and Fig. 48) (1932).

37.   M. Kasha and D. E. Brabham in *Singlet Oxygen,* edited by H. H. Wasserman and R. W. Murray (Academic, New York, 1979), Chapter 1.

38.   B. J. Moss, F. W. Bobrowicz, and W. A. Goddard, *J. Chem. Phys.* **63**, 4632 (1975).

39.   E. Hückel, *Z. Physik* **60**, 423 (1930).

40.   D. R. Kearns, *J. Am. Chem. Soc.* **91**, 6554 (1969).

41.   L. Salem, *Pure Appl. Chem.* **33**, 317 (1973).

42.   L. B. Harding and W. A. Goddard, *J. Am. Chem. Soc.* **102**, 439 (1980).

43.   S. D. Peyerimhoff and R. J. Buenker, *Chem. Phys. Lett.* **16**, 235 (1972); C. F. Bender and E. R. Davidson, *J. Phys. Chem.* **70**, 2675 (1966).

44.   J. T. Vanderslice, E. A. Mason, and W. G. Maisch, *J. Chem. Phys.* **32**, 515 (1960); G. Herzberg, *Molecular Spectra and Molecular Structure* (Van Nostrand, Princeton, N. J., 1966) p. 560.

45.   N. J. Turro, *Modern Molecular Photochemistry* (Benjamin, New York, 1978), equation (14.12).

46.   R. S. Mulliken, *Phys. Rev.* **41**, 751 (1932).

47.   R. J. Buenker and S. D. Peyerimhoff, *Chem. Phys.* **9**, 75 (1976).

48.   B. Brooks and H. F. Schaefer, *J. Am. Chem. Soc.* **101**, 307 (1979).

49.   F. Hund in *Linienspektren und periodisches System der Elemente* (Springer, Berlin, 1927), p. 124.

50.   H. Kollmar and V. Staemmler, *Theoret. Chim. Acta (Berl.)* **48**, 223 (1978).

51.   J. Michl, *J. Am. Chem. Soc.* **98**, 6427 (1976); *Pure Appl. Chem.* **41**, 507 (1975).

52.   U. Kaldor and I. Shavitt, *J. Chem. Phys.* **48**, 191 (1968).

53.   A. J. Merer and R. S. Mulliken, *Chem. Rev.* **69**, 639 (1969).

54.   R. J. Buenker, S. D. Peyerimhoff, and H. L. Hsu, *Chem. Phys. Lett.* **11**, 65 (1971); S. D. Peyerimhoff and R. J. Buenker, *Theoret. Chim. Acta (Berl.)* **27**, 243 (1972); R. J. Buenker, V. Bonacic-Koutecky, and L. Pogliani, *J. Chem. Phys.* **73**, 1836 (1980).

55.   G. Orlandi, P. Palmieri, and G. Poggi, *J. Am. Chem. Soc.* **101**, 3492 (1979); G. Orlandi and W. Siebrand, *Chem. Phys. Lett.* **30**, 352 (1975).

56.   W. A. Goddard, T. H. Dunning, W. J. Hunt, and P. J. Hay, *Acc. Chem. Res.* **6**, 368 (1973).

57.   P. J. Hay, T. H. Dunning, and W. A. Goddard, *J. Chem. Phys.* **62**, 3912 (1975); D. Grimbert and A. Devaquet, *Molec. Phys.* **27**, 831 (1974).

58.   J. S. Wright, *Can. J. Chem.* **51**, 139 (1973).

59.   B. S. Rabinovitch, E. W. Schlag, and K. B. Wiberg, *J. Chem. Phys.* **28**, 504 (1958); S. W. Benson, ibid. **34**, 521 (1961); S. W. Benson and P. S. Nangial, ibid. **38**, 18 (1963).

60.   R. Hoffmann, *Acc. Chem. Res.* **4**, 1 (1971).

61.   Y. Jean, L. Salem, J. S. Wright, J. A. Horsley, C. Moser, and R. M. Stevens, *Pure Appl. Chem. Suppl.* (23rd Congr.) **1**, 197 (1971); Y. Jean, D. Sc. thesis, Université de Paris-Sud, Orsay, France, 1973.

62.   R. G. Bergman in *Free Radicals,* Vol. 1, edited by J. K. Kochi (Wiley, New York, 1973), p. 191.

63.   J. A. Berson, L. D. Pedersen, and B. K. Carpentier, *J. Am. Chem. Soc.* **98**, 122 (1976).

64.   E. F. Hayes and A. K. Q. Siu, *J. Am. Chem. Soc.* **93**, 2090 (1971).

65.   W. Adam, *Acc. Chem. Res.* **12**, 390 (1979).

66.   E. J. York, W. Dittmar, J. R. Stevenson, and R. G. Bergman, *J. Am. Chem. Soc.* **94**, 2882 (1972).

67.   R. S. Hutton, M. L. Manion, H. D. Roth, and E. Wasserman, *J. Am. Chem. Soc.* **96**, 4680 (1974).

68.   G. E. Palmer, J. R. Bolton, and D. R. Arnold, *J. Am. Chem. Soc.* **96**, 3708 (1974).

69.   J. H. Davis, W. A. Goddard, and R. G. Bergman, *J. Am. Chem. Soc.* **98**, 4015 (1976).

70. G. Levin and W. A. Goddard, *J. Am. Chem. Soc.* **97,** 1649 (1975).
71. L. Salem, *The Molecular Orbital Theory of Conjugated Systems* (Benjamin, New York, 1966).
72. J. Paldus and A. Veillard, *Molec. Phys.* **35,** 445 (1978).
73. J. Paldus and J. Cizek, *J. Chem. Phys.* **54,** 2293 (1971); J. M. McKelvey and G. Berthier, *Chem. Phys. Lett.* **41,** 476 (2976) and Section 3.1.
74. W. A. Goddard, *J. Am. Chem. Soc.* **94,** 793 (1972), discussion of Fig. 2; K. Yamaguchi, Y. Yoshioka, and T. Fueno, *Chem. Phys.* **20,** 171 (1977); L. Salem, *Nouv. J. Chim.* **2,** 559 (1978); Y. Jean, unpublished calculations.
75. W. T. Borden and E. R. Davidson, *Ann. Rev. Phys. Chem.* **30,** 125 (1979).
76. C. Y. Lin and A. Krantz, *Chem. Commun.* 1111 (1972); O. L. Chapman, C. L. McIntosh, and J. Pacansky, *J. Am. Chem. Soc.* **95,** 614 (1973).
77. W. T. Borden, *J. Am. Chem. Soc.* **97,** 5968 (1975).
78. H. Kollmar and V. Staemmler, *J. Am. Chem. Soc.* **99,** 3583 (1977).
79. W. T. Borden, E. R. Davidson, and P. Hart, *J. Am. Chem. Soc.* **100,** 388 (1978).
80. J. A. Jafri and M. D. Newton, *J. Am. Chem. Soc.* **100,** 5012 (1978).
81. W. von E. Doering and L. Birladeanu, *Tetrahedron* **29,** 449 (1973).
82. J. H. Davis and W. A. Goddard, *J. Am. Chem. Soc.* **98,** 303 (1976).
83. W. T. Borden and L. Salem, *J. Am. Chem. Soc.* **95,** 932 (1973).
84. W. T. Borden and E. R. Davidson, *J. Am. Chem. Soc.* **99,** 4587 (1977).
85. W. T. Borden, *J. Am. Chem. Soc.* **97,** 2906 (1975); E. R. Davidson and W. T. Borden, *J. Am. Chem. Soc.* **99,** 2053 (1977).
86. D. R. Yarkony and H. F. Schaefer, *J. Am. Chem. Soc.* **96,** 3754 (1974).
87. D. M. Hood, H. F. Schaefer, and R. M. Pitzer, *J. Am. Chem. Soc.* **100,** 8009 (1978).
88. D. A. Dixon, R. Foster, T. A. Halgren, and W. N. Lipscomb, *J. Am. Chem. Soc.* **100,** 1359 (1978).
89. J. A. Berson, *Acc. Chem. Res.* **11,** 446 (1978); M. S. Platz and J. A. Berson, *J. Am. Chem. Soc.* **99,** 5178 (1977).

4

Electronic Control by Orbital Symmetry

Chemistry, a few times each century, undergoes revolutions in thinking, which are the logical consequence of the patient groundwork and foundations laid during previous decades. The discovery by Woodward and Hoffmann of "orbital symmetry conservation" was a revolution along these very lines. It was a major breakthrough in the field of chemical reactions in which notions preexisting in other fields (orbital correlations by Mulliken, and nodal properties of orbitals in conjugated systems, by Coulson and Longuet-Higgins) were applied with great conceptual brilliance to a far-reaching problem. Chemical reactions were suddenly adorned with novel significance. In this chapter we discuss the basic foundations of orbital symmetry control in chemical reactions.*

4.1 Two Helium Atoms Approaching Each Other

Consider two helium atoms, each with a pair of electrons in a $1s$ orbital, approaching each other. The conventional orbital interaction diagram shows that a bonding $1s\sigma_g$ molecular orbital and an antibonding $1s\sigma_u$ orbital are formed, each with a pair of electrons:

$$(4\text{-}1)$$

The two atoms repel each other. A traditional explanation for this repulsion is that the antibonding electrons are more antibonding than the bonding electrons are bonding. This statement is justified by the relative orbital energies of $1s\sigma_g$ (proportional to $H_{AB}/(1 + S_{AB})$, where H_{AB} is the matrix element of some effective one-electron Hamiltonian and S_{AB} is the overlap between $1s_A$ and $1s_B$) and of $1s\sigma_u$

*For a major overview of material contained in Chapters 4–6, see Ref. 1.

(proportional to $H_{AB}/(1 - S_{AB})$). A *second* explanation is to invoke the fact that all four electrons tend to be in the same region of space—in the atomic overlap region $\phi_A\phi_B$ between the atoms—and that this is forbidden by Pauli's exclusion principle for those pairs of electrons with parallel spin. Even if this statement is modulated by saying that the bonding electrons occupy the overlap region to a greater extent than the antibonding electrons, the simultaneous occupancy is sufficient to cause repulsion.

A *third* explanation would draw out the orbital behavior along the full reaction path by *correlating* the orbitals in the reactant with those in the ultimate product: those of a beryllium atom (we assume formally that the helium nuclei can be appropriately fused into a beryllium nucleus). Figure 4-1 shows such a *correlation diagram*.[2,3] The molecular orbitals of the helium pair are placed on the left in order of increasing energy and similarly on the right for the atomic orbitals of beryllium. Common symmetry properties (local orbital angular momentum of the helium molecular orbitals around the C_∞ infinite-fold molecular axis and parity with respect to the center of inversion) are then used to join the levels on the left with those on the right. Energy is an additional discriminating factor and, for a given starting level, one works upward in energy in the product levels until the first level of same symmetry is found. The $1s\sigma_g$ bonding orbital of He_2, with no nodal surfaces, goes smoothly into the $1s$ orbital of beryllium. The $1s\sigma_u$ antibonding orbital, with its nodal plane bisecting the internuclear axis, correlates, not with the $2s$ orbital of beryllium (which has a *radial* nodal surface and g symmetry), but with the $2p$ orbital of beryllium that points along the axis. This orbital has such high energy that the approach of the two atoms is energetically prohibitive. Of course, the $1s$ orbital of beryllium has a very low energy, but in the present case this is offset by the nearly infinite Coulomb repulsion of the two helium nuclei that must be included in the total energy.

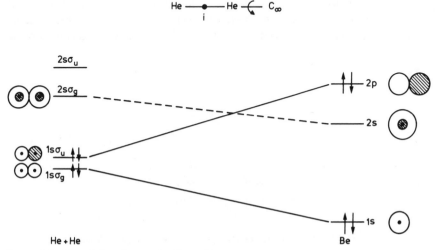

FIGURE 4-1.   Correlation diagram for He + He → Be. Controlling symmetry elements are shown above.

The correlation diagram thus shows that the pair of ground helium atoms "correlates" with a *doubly excited* beryllium atom. To yield the *ground* beryllium atom, one would have to start with a helium pair in which two electrons were excited from $1s\sigma_u$ into the molecular orbital that correlates with $2s_{Be}$. This is the next bonding molecular orbital, $2s\sigma_g$ (Fig. 4-1, dotted lines), formed by in-phase combination of the $2s$ orbitals of the helium atoms.

An important caveat is to ensure that the present description, obtained by drawing out one–electron energy levels obtained from some one-electron Hamiltonian (Hartree-Fock or other), remains valid when the instantaneous Coulombic repulsion of the electrons is included (even in the Hartree-Fock method, this repulsion is included only in an average manner). The extent to which electron correlation "spoils" such one-electron diagrams is discussed in Chapter 5.

4.2 A "Forbidden" Reaction: Face-On Dimerization of Two Ethylene Molecules

We now set out to construct one of the Woodward-Hoffmann correlation diagrams,* that for the face-on dimerization of two ethylene molecules[6]:

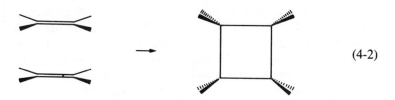

$$(4\text{-}2)$$

This diagram bears similarities with that in Section 4.1. Instead of a C_∞ axis and center of inversion as symmetry elements, there are two planes. One of them, P_{intra}, is a local symmetry plane common to both molecules (Fig. 4-2). The second one, P_{inter}, lies midway between the molecular planes and plays somewhat the same role as the inversion center previously. This second plane allows molecular orbitals

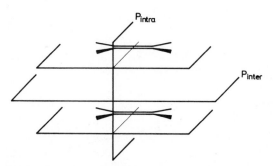

FIGURE 4-2.   Face-on (or parallel) approach of two ethylene molecules.

*See Woodward and Hoffmann's 1969 and 1970 works.[4] Their first epoch-making work was in 1965.[5]

to be built for the *pair* of molecules, by combining the local molecular orbitals. The symmetry—symmetric S or antisymmetric A—relative to P_{inter} determines the nonexistence or existence of a nodal plane between the ethylene molecules and parallel to them. The symmetry relative to P_{intra} determines which *local* orbitals are legitimate for combination: only S orbitals together and only A orbitals together.

As shown in Fig. 4-3, the local π bonding orbitals that have S symmetry (P_{intra}) combine to form a dimer S pair and a dimer A pair relative to P_{inter}. Similarly, the π^* orbitals, with local A symmetry (P_{intra}), form an S intermolecular pair and an A intermolecular pair. On the cyclobutane product side consideration is restricted to the σ and σ^* CC bond orbitals in the two newly formed bonds. This time the local symmetry is determined relative to P_{inter} since this plane cuts through the bonds. This symmetry is S for the σ_{CC} orbitals and A for the σ^*_{CC} ones. Formation of combinations of these bond orbitals now gives symmetries relative to P_{intra}.

As each orbital has two symmetry labels, it is possible to correlate the four reactant orbitals with the four product orbitals; these correlations are shown in Fig. 4-3. It should be noted that a *single symmetry label*—whether local symmetry of reactants ("interbond" symmetry of products) or "intermolecular" symmetry of reactants (local symmetry of products)—would suffice to obtain these same correlations. In either case the pair of ground ethylene molecules correlates with a

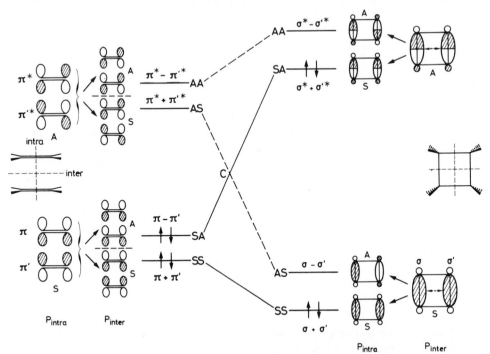

FIGURE 4-3.   Woodward-Hoffmann orbital correlation diagram for the face-on dimerization of ethylene. Orbital symmetries are relative respectively to P_{intra} and P_{inter}. Primes refer to the second ethylene molecule or to the second CC σ bond.

BELMONT COLLEGE LIBRARY

doubly excited ($\sigma \to \sigma^{*2}$) cyclobutane molecule. The result is identical with that for the pair of helium atoms.

In view of the steep uphill slope imposed on one electron pair, the reaction can be said to be *"forbidden"* by orbital symmetry. For the reaction to occur, the electron pair would have to rise in energy until the crossing point C between rising SA orbital and descending AS orbital, at which point the reaction can legitimately switch from one orbital to the other. However, C lies midway between bonding and antibonding orbitals, at the nonbonding level. Hence the energy of two bonding π electrons—the equivalent of one double bond, or 270 kJ/mole—will have been spent to reach the transition state. This is prohibitive, and the reaction does not occur in this "concerted" manner, insofar as the steep upward slope is carried over from the *orbital* one-electron picture to the *state* picture in which electron motions are correlated (see Chapter 5).

In the absence of a symmetry-allowed pathway (for two olefins, there is an alternative allowed pathway[4] but that is strongly sterically hindered) the reaction is expected to occur in *two steps* (see Fig. 2-3*b*). This stepwise pathway leads to a diradical (which may or may not be an intermediate; see Section 2.1) that should be preferred over the forbidden pathway leading to C. Indeed, granted that two electrons become nonbonding in both pathways, at least in the diradical there is compensating creation of one bond. In the present case passage of the reverse reaction, the pyrolysis of cyclobutane, through the tetramethylene diradical,

$$(4\text{-}3)$$

apparently in a number of possible conformations, seems to have been confirmed experimentally.[7]

It is remarkable that the transition state, with a ring of four orbitals interacting σ-wise:

$$(4\text{-}4)$$

bears some resemblance to an *antiaromatic* molecule such as cyclobutadiene (Section 3.9). This was pointed out by Dewar and Zimmerman in particular.[8,9] Although there is a nodal surface on each atom as one goes around the ring, each interatomic overlap is positive as for π conjugation in cyclobutadiene. The inherent instability of the latter, as a result of its diradical character, is thus carried over into the transition state C here (Section 4.6).*

*The aromatic nature of the transition state for the Diels-Alder addition (Section 4.3) was pointed out as early as 1939 by Evans.[10] However, contrary to later assertions, the crucial Woodward-Hoffmann alternation of "allowedness" and "forbiddenness" of cycloadditions as electron pairs are added to the system is nowhere to be found in Evans' work. The two following letters illustrate the debate which arose after the Woodward-Hoffman discovery. The author thanks Michael Dewar for permission to reproduce his letter.

Professor R.B. WOODWARD
Department of Chemistry
Harvard University
12, Oxford Street
Cambridge, Mass. 02138
U.S.A.

January 25, 1971

Dear Professor Woodward,

I have carefully read and re-read M.G. Evans' article on "the Activation Energies of Reactions involving conjugated systems" (Trans. Far. Soc. **35,** 824 (1939)). In this article Evans makes very clear that the delocalization of electrons in the transition state can provide for a special stabilization ("resonance energy"). It is also quite clear that in his mind this special stabilization operates both for cyclic and linear transition states (see bottom of p. 831: ". . . in the transition state (the mobile electrons) simulate the behaviour in a benzene structure. A similar argument can be applied to the reactions by the end on method attack, where in the transition state the mobile electrons will tend to behave like those in hexatriene"), but more strongly in the former: "the lowering of the activation energy due to the resonance effect will be greater in cyclisation reactions than in chain formation" (p. 832).

Nowhere can I find any distinction between cyclic transition states with different numbers of carbon atoms or any reference to the essential ALTERNATION in stability as a function of the number of atoms which you discovered with Roald Hoffmann. An extreme view would be to contend that this alternation is *implicit* in Evans' statement that the cyclic transition states are strongly stabilized by resonance energy (since automatically then transition-state stabilization energies should follow the rules of aromaticity). However this all-important alternation of properties did not occur to Evans, since its immediate consequence is precisely the preference for a two-step "linear" attack in these cases where the cyclic stabilization is poor—a consequence which is certainly at variance with his previously quoted statement of p. 832.* Such an important deduction was probably not possible within the state of Chemistry in 1939.

I honestly conclude that the basic principle of alternating "forbiddenness" and "allowedness" in concerted reactions cannot be found, either explicitly or implicitly, in Evans' paper. The only matter which I cannot, of course, guarantee in a scientific manner is my own impartiality.

Sincerely yours,

L. Salem

*Note that Evans' statement p. 832 follows from his argument: "The energy levels of the mobile electrons lie lower in cyclical structures than in straight-chain compounds with the same number of centres available." This is wrong in the 4n case (see for instance n = 1 or n = 2). Were it correct, all conjugated rings would be aromatic and there would be no alternation of properties.

Copies: Professor Roald Hoffmann
Professor Michael Dewar

The University of Texas at Austin
AUSTIN, TEXAS 78712

Department of Chemistry   February 2, 1971

Professor Lionel Salem
Laboratoire de Chimie Théorique
Batiment 496
Faculté des Sciences
91-Orsay, France

Dear Lionel,

Many thanks for the copy of the letter you sent to Bob Woodward.

Frankly it seems to me a bit irrelevant. Of course Evans could not have predicted the alternation in "allowed" pericyclic reactions with ring size; after all there was no real reason for believing that the 4n-electron cyclic polyenes would be destabilized, i.e. antiaromatic, prior to my 1952 JACS papers (where the term "antiaromatic" was first introduced). This, however, is not the point; what Evans did was to point out the topological relationship between the delocalized MOs in transition states of what Woodward and Hoffmann call pericyclic reactions and the π MOs of cyclic polyenes. This to my mind was at the time an extraordinary insight and together with my PMO treatment of aromaticity, and Craig's recognition in 1957 of the existence of antiHückel systems, it provided a complete theory of reactions of this type, waiting for use when they should be discovered.

This to my mind is the key point. Until reactions of this type *were* discovered, it would have been pointless speculation to predict their behaviour on the basis of qualitative theories. The difference between 1938 and 1964 was that in the meantime not only had the theory of aromaticity been developed, but also a whole spectrum of new reactions had been discovered due to the development of g.l.c. and the consequent interest in thermal reactions of hydrocarbons.

This is not to say that reactions of this kind were not known and interpreted before; certainly I have been including in my lecture course since the mid-fifties a discussion of what I then called "reactions involving cyclic transition states", using as examples, among others, the Diels-Alder reaction, the Claisen rearrangement, the ene reaction, addition of diazomethane to olefines, and the pyrolysis of ricinoleic acid. I might add that on a visit to Du Pont in 1958, when asked why certain olefines gave cyclobutene derivatives with dienes rather than "normal" Diels-Alder products, I replied at once that the reactions must involve two–step processes via intermediate biradicals since the concerted cycloaddition to a 4-ring would involve an antiaromatic transition state, while cyclisation of a biradical to a 4-ring rather than a 6-ring would be favoured from considerations of entropy. Again, my first exposure to the original Woodward-Hoffmann paper was at one of my research seminars since my issue of JACS arrived later than the one in our library; I at once explained the conrotatory ring opening of cyclobutene in terms of an aromatic antiHückel transition state. All these ideas seemed to me obvious deductions from the literature and consequently not worth publishing; indeed until quite recently, no journal would have published suggestions of this kind, as I know from early experience! New theories and new experiments were publishable but not interpretations of other peoples' work in terms of existing theory. Indeed, I was told later that in 1952 JACS had a set policy against publication of *any* theoretical papers on organic chemistry and that publication of mine created a precedent. I only published the treatment of pericyclic reactions in terms of what I like to call Evans' principle, partly because I think it was a very seminal suggestion and partly because he was a personal friend of mine, when it became clear that a lot of ideas which I had assumed to be familiar to others in fact were not.

Best wishes,

Yours sincerely,

Michael J.S. Dewar

MJSD:pc

Copies: Professor R.B. Woodward
Professor R. Hoffmann

4.3   An "Allowed" Reaction: Diels-Alder Cycloaddition of Ethylene and Butadiene

Our next example considers the approach of two different molecules, butadiene and ethylene, again with their planes parallel:

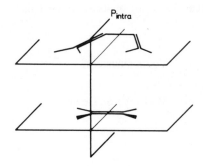

$$(4\text{-}5)$$

The problem is the same as if we were to combine a helium atom with a neon atom: the C_∞ axis that determines the local orbital angular momentum of the atomic orbitals would subsist, but the inversion center, which determines symmetries of the combined atomic orbitals, would disappear. Here the intermolecular symmetry plane P_{inter} disappears, and only a common plane P_{intra} that determines the local symmetry of the π molecular orbitals of the two molecules subsists (Fig. 4-4).

Relative to this plane, the familiar amplitude distribution of the Hückel π molecular orbitals of butadiene and ethylene[11] immediately yields their symmetry. The butadiene orbitals, starting from the deepest bonding one, are alternatively S and A, whereas the π and π^* orbitals of ethylene are respectively S and A. The product orbitals that must be considered are the π and π^* orbitals of the newly formed double bond of cyclohexene (former single bond of butadiene) and the σ_{CC}, σ_{CC}^* orbitals of the two new CC single bonds of cyclohexene. Interestingly, although the π, π^* pair have a clear-cut symmetry *(S, A)* relative to P_{intra}, the σ orbitals must be combined to obtain appropriate symmetries, as in the cyclobutane case (Fig. 4-5).

The orbital correlations between reactant and product are then drawn out in a straightforward manner. Starting from the lowest energy levels in reactant, we correlate the lowest S orbital of reactant with the lowest S orbital of product and so on. Contrary to the previous case, all bonding orbitals of reactant correlate with

FIGURE 4-4.   Controlling symmetry plane in the Diels-Alder addition of ethylene to butadiene.

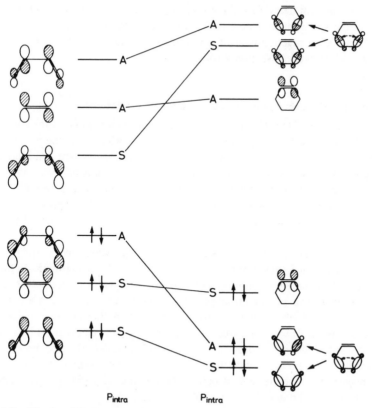

FIGURE 4-5.   Woodward-Hoffmann orbital correlation diagram for the Diels-Alder cycloaddition of butadiene to ethylene (see Fig. 4-12).

bonding orbitals of product. No electron pair is raised up to the nonbonding level, the reaction is "symmetry allowed" and the bonds can form in a "*concerted*" manner. It should be noted, however, that this allowedness does not automatically imply that the reaction will be facile. In fact, the activation energy for cycloaddition of ethylene to butadiene is quite large (144 kJ/mole).* Yet none of the three occupied orbitals appears to rise in the correlation diagram (Fig. 4-5). The answer to this dilemma is that in the initial steps of the reaction, strain energy must be spent to stretch the three double bonds (which are all destined to become single bonds). This "deformation energy"[13] affects essentially the σ-orbital levels of reactants, which are not a party to the correlation diagram. The σ orbitals for the three C=C bonds will rise at the outset. This is also true to a lesser extent for the π orbitals, in particular for the two more labile ones (see Section 4.4 and Fig. 4-7)—a feature not revealed by the straight-line correlation in Fig. 4-5.

*This value is obtained from the experimental enthalpy of formation of product at 0°K (− 135 kJ/mole) and from the activation energy of 279 kJ/mole for the reverse Diels-Alder reaction of cyclohexene.[12]

Just as the symmetry forbiddenness of the dimerization of ethylene can be related to the antiaromatic nature of the transition state, the symmetry allowedness of the Diels-Alder reaction is justifiably linked to its aromaticlike, benzene-resembling transition state[8,9] (Section 4.6).

4.4 How the Orbitals Really Correlate: Natural Correlations

A striking feature of the Woodward-Hoffmann correlation diagram for Diels-Alder cycloaddition is the correlation of initial ethylene π orbital with the newly created π orbital at the opposite end of the cyclohexene ring! Woodward and Hoffmann[14] went to great pains to show that this "strange" correlation is perfectly understandable since the ethylene π orbital mixes with the two butadiene S orbitals (one full and one empty) as the reaction proceeds. This mixture will progressively kill the amplitude on the ethylene double bond while creating it on the central butadiene bond.

Millie[15] was first to wonder, however, whether one should not correlate S (ethylene) directly with S (newly formed σ_{CC} bonds), which has amplitude on the same atoms, and S (butadiene) with cyclohexene double bond S orbital—that is, pairs of orbitals that have maximum amplitude on the same atoms. Such a correlation is shown in Fig. 4-6b. A major problem, however, is that this correlation forces the two S orbitals to *intersect*. Such an intersection, between two ortibals of same symmetry is forbidden.[16] Indeed, at intersection I (Fig. 4-6) the two orbitals S_1 and S_2 have the same energy. Since they also have the same symmetry, they can *mix* through whatever one-electron effective Hamiltonian H_{eff} (Section 1.8) has been used to draw out the diagram. The matrix element is

$$\langle S_1 \mid H_{eff} \mid S_2 \rangle \neq 0 \qquad (4\text{-}6)$$

unless there is a fluke cancellation due to geometric factors. In practice, such cancellations have never been found, and orbitals with the same symmetry quantum numbers cannot cross on a one-dimensional energy diagram (see Section 5.9).[16] Therefore, the Millie correlation is at least formally incorrect.[17]

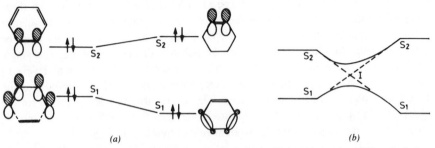

FIGURE 4-6. Correlation between the bonding S orbitals in the Diels-Alder cycloaddition of ethylene to butadiene (for intersection I and dotted lines, see discussion in text): *(a)* Woodward-Hoffmann[4]; *(b)* Millie.[15]

It might happen, however, that the *intended* correlations (dotted lines, Fig. 4-6*b*) between ethylene π and cyclohexene σ_{CC} and between butadiene π and cyclohexene π remain, with simply an "avoided" intersection reflecting the orbital mixing (see Section 5.8). The correlations would then follow the full lines of Fig. 4-6*b*. However, it appears that there is no such "memory" of intended correlations in the form of a near crossing in the Diels-Alder reaction. The correct, calculated correlations[18] are shown in Fig. 4-7. The two *S* orbitals remain well separated throughout the reaction path, possibly because reactant strain (stretch of double bond in ethylene that increases the energy of π ethylene) and equally the strain on product side (stretch of ring double bond affecting π cyclohexene if reaction path is taken backward) seem to offset any intended correlation.

Despite its apparent failure in the present case,* the concept that orbitals should correlate in a natural manner is a powerful one. Introduced by Devaquet, Sevin, and Bigot[19] "natural correlations" require that orbitals remain localized on the same atoms as far as possible and that the phase properties of the correlating orbitals (position and nature of the nodal surfaces, see also Section 4.5) remain the same as far as possible. An example of such natural correlations is shown in Fig. 4-8 for the *n*-initiated hydrogen capture in a simple formaldehyde–methane model abstraction reaction. From a purely energetic point of view, the high-lying lone-pair *n* orbital on oxygen would tend to be correlated with the odd orbital on the methyl radical. But its natural correlation is with the deeper σ_{OH} orbital, whereas n_{CH_3} originates with the σ_{CH}-bond orbital (Fig. 4-8, dotted lines). Hence the naturally correlated orbitals intersect. The avoided intersection still leaves a memory as the orbital energies approach and then separate (Fig. 4-8, solid lines). This memory itself has an important observable trace since it determines the activation energy for the photochemical n,π^* abstraction reaction.[19]

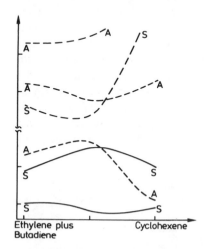

FIGURE 4-7. Actual calculated one-electron correlations[18] for the π levels in the Diels-Alder reaction. Reprinted with permission from R. E. Townshend, G. Ramunni, G. Segal, W. J. Hehre, and L. Salem, *J. Am. Chem. Soc.* **98**, 2190 (1976). Copyright 1976 by American Chemical Society.

Ethylene plus Butadiene — Cyclohexene

*A natural correlation diagram that leads correctly to Fig. 4-7 can be drawn out if the ethylene and butadiene orbitals are first combined relative to the pseudosymmetry plane between the molecules (B. Bigot, private communication).

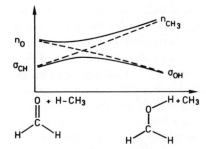

FIGURE 4-8.　Natural correlations for hydrogen abstraction from methane by formaldehyde.[19] Symmetry–controlling element is molecular plane of ketone. Dotted lines, intended correlation; full lines, final correlation.

4.5　Orbital Symmetry Allowedness, Forbiddenness, and Nodal Surfaces

Let us try to delve further into the origin of the forbiddenness observed in the approach of two helium atoms or in the face-on dimerization of ethylene. In the first case a $1s\sigma_u$ orbital with a nodal plane cutting through the molecular axis must transform into a $2s\sigma_g$ orbital that has also a nodal surface, but a "radial" one surrounding the two nuclei (Fig. 4-9a). By "must transform" we mean that this is the requirement that the nodal properties of the wave function must satisfy if the highest electron pair, originally in $1s\sigma_u$, is to follow the lowest-energy path and finish in $2s\sigma_g$. Similarly, the top-bonding electron pair in the ethylene "dimer" must arrange to exchange a horizontal nodal plane, between the two molecules, for a vertical plane, separating the two bonds (Fig. 4-9b).

In each case the number of nodal surfaces remains constant; hence theorems relating the number of nodal surfaces (and nodal regions that they separate) to energy ordering[20] cannot be used directly. The ordering of energy levels is changing here between reactant and product solely because of the inherent change in *shape* of the nodal surface, itself triggered by a change in *shape* of the molecular system. This is best illustrated by showing the squeezing reaction of a rectangular box. Figure 4-10 gives the nature of the three lowest "particle-in-a-box" levels for such a reaction. The correlation diagram, preserving symmetry or antisymmetry with respect to horizontal and vertical symmetry planes, is identical with those of the two symmetry-forbidden reactions (Figs. 4-1 and 4-3). Here the *kinetic energy* of the particles is the controlling factor. In the horizontally shaped box the lowest pair adopts a sine wave filling the entire box, but the second pair, whose function must be orthogonal to the first, is obliged to divide its amplitude and density into two

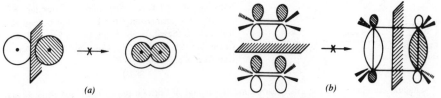

FIGURE 4-9.　Changes of nodal surfaces required of the top bonding electron pair: *(a)* in the approach of two helium atoms; *(b)* in the face-on dimerization of ethylene.

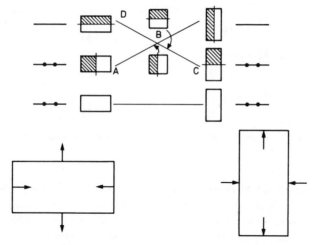

FIGURE 4-10.   Squaring of a rectangular box. Particle-in-a-box levels and nodal properties. Levels for $a = \sqrt{2b}$. Correlations are illustrated for four particles.

parts separated by a vertical nodal plane. Its kinetic energy increases, but would be even higher if it were trapped in the third orbital with two thin slabs separated by a horizontal nodal plane, with the large kinetic energy for vertical motion finding no compensation in the low kinetic energy for horizontal motion (Fig. 4-11b). The lower energy of the second orbital is due to the fact that the particle kinetic energy is "average" there in both directions (Fig. 4-11a). As the reaction proceeds, however, the electron pair must switch from this orbital to the other one.

Hence the "symmetry-forbiddenness" of Woodward-Hoffmann–forbidden reactions can be ascribed in this model to a rise in electronic kinetic energy as the molecular shape changes so that two of the four electrons become squeezed along one of their directions of motion (path AB in Figure 4-10). Only after the transition state is reached can the squeezing be relieved (path BC) by the change in nodal surface, with a concomitant required *change in preferred direction of motion* of the electron pair.

If the number of particles is six instead of four, the situation changes. While one pair is squeezed (AB), another is relieved (DB) and decreases its kinetic energy;

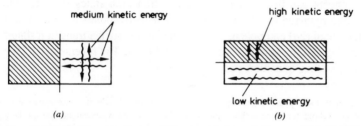

FIGURE 4-11.   Particle kinetic energy for two different nodal planes: *(a)* lower orbital; *(b)* higher orbital.

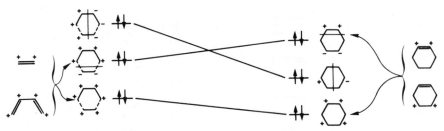

FIGURE 4-12.   Schematic changes in the nodal surfaces of the top bonding orbitals during the Diels-Alder reaction (see Fig. 4-5).

the situations are reversed in the second half of the reaction. The overall kinetic energy remains approximately constant. This is the case for Woodward-Hoffmann "symmetry-allowed" reactions such as the Diels-Alder reaction. Returning to Fig. 4-5, we see that the top pair has a vertical nodal surface whereas the middle pair, although without nodal surface at the start, soon acquires a horizontal one through the out-of-phase mixing between ethylene S orbital and the S orbital of same symmetry on butadiene (Fig. 4-12). In the product the middle pair has acquired the vertical plane while the top pair has a horizontal one.*

Under these circumstances it is not surprising that orbital-symmetry-controlled reactions can also be analyzed in terms of requirements involving continuous-phase relationship changes between GVB orbitals[21] or by correlating localized orbitals.[22] A purely valence-bond deduction of Woodward-Hoffmann rules also exists.[23]

4.6   Allowedness, Forbiddenness, and Aromatic and Antiaromatic Transition States

As mentioned briefly in Sections 4.2 and 4.3, the symmetry allowedness or forbiddenness of reactions can be related to the orbital properties of the transition state. For the two cycloadditions considered earlier, the Dewar-Zimmerman approach[8,9] uses as model transition state the conjugated ring with the same number of carbon centers. This is true for any Woodward-Hoffmann "pericyclic" reaction[4] in which there are only "suprafacial"[4] segments (or eventually, an even number of "antarafacial" segments). Figure 4-13 is a reminder of the orbital amplitudes for the labile electrons in the four-atom, four-electron antiaromatic transition state (Fig. 3-7) and in the six-atom six-electron aromatic transition state.

The concept can be extended to include transition states in which the atomic orbitals are organized in a Möbius strip.[24] This corresponds to rings of atoms with one antarafacial ($2a$) component. For four electrons, such a transition state becomes fully bonding and the reaction is symmetry allowed. The reverse occurs for the six-electron, six-atom transition state (Fig. 4-14).

---

*This horizontal nodal plane in the top bonding product orbital, which is not apparent in Fig. 4-5, appears if one allows for some residual interaction between ring double bond and the newly formed ring σ bonds.

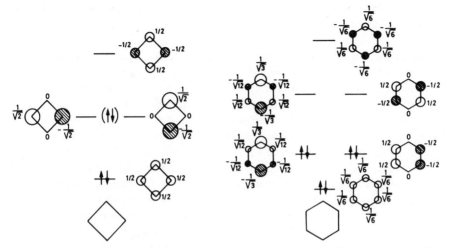

FIGURE 4-13. Orbital amplitude patterns for antiaromatic and aromatic transition states (see also Fig. 3-7). See Table 3-11 for the electron distribution in the nonbonding orbitals.

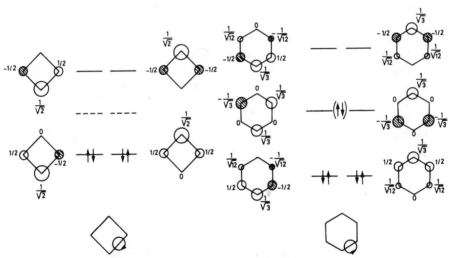

FIGURE 4-14. Orbital amplitude patterns for aromatic and antiaromatic "Möbius"[24] transition states. The Möbius nature of the ring is indicated by the arrow. The author is grateful to E. Heilbronner for the Möbius orbital coefficients.

111

Just as the nodal properties along the reaction path are related to the kinetic behavior of the electron pairs (Section 4.5) the antiaromatic or aromatic nature of the transition state may also be translated into properties of electron pairs. For instance, the pair probability distribution—defined, say, by calculating the simultaneous probability of finding two electrons of opposite spin at the same point in space—differs strongly from one type of transition state to the other.[25] For the aromatic transition state, the pair probability distribution for the four top bonding electrons is a smooth positive function, whereas for the antiaromatic transition state, the distribution for the two top electrons is split into four parts, with two diagonal planes, where it vanishes (Fig. 4-15). In the first case the electron pairs can be considered as "preserved," and in the second case the top pair can be said to be "broken"—in the sense that the two electrons of opposite spin tend to be simultaneously in different parts of the square. This is indeed what is suggested by the wave function for the lowest singlet of cyclobutadiene, particularly when the lozenged pair of nonbonding orbitals is used for this singlet (Table 3-11). This "broken" pair nature of the wave function agrees with the nonbonding character of the top two electrons in the Woodward-Hoffmann correlation diagram. It can also be related to the kinetic-energy analysis of the previous section by saying that the *required change in direction of motion* of the electrons (from low kinetic energy along the vertical direction to low kinetic energy along the horizontal direction; at B, Fig. 4-10) pulls the pair apart.

The use of transition-state structures to determine symmetry-allowedness or forbiddenness—despite the sometimes entangled and highly noncoplanar nature of real-life transition structures, which complicates their analogy with simple planar or Möbius rings—is a powerful tool. It reveals the deep analogy between the Woodward-Hoffmann rules and Hückel's $(4n + 2)$-rule.[11] The great stability of conjugated planar ring systems with *odd* numbers of electron pairs can thus be extended to the field of orbital-symmetry-controlled reactions.[26] Note finally that the favorable nature of odd-numbered-pair systems is true also for the first three rare gas atoms: helium (one pair), neon (five pairs), and argon (nine pairs) because of the nodal properties of spherical harmonics.

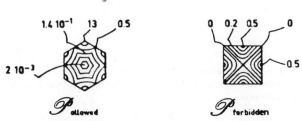

FIGURE 4-15. Rough sketches of the pair probabilities[25] for the labile electrons in allowed (four electrons) and forbidden (two electrons) pericyclic transition states. In $\mathscr{P}_{forbidden}$ the pair probability vanishes along the diagonals and is positive elsewhere. Numbers are in units of a^4, where a^4 is the value of the squared atomic density Φ_A^4 at the middle of the bonds leading from A. Reprinted with permission from L. Salem, *Nouv. J. Chimie* **2,** 559 (1978). Copyright 1978 by CNRS-Gauthier Villars.

4.7 Effect of a Metal Atom on an Orbital-Symmetry-Forbidden Reaction

In the early days of orbital symmetry theory metal catalysis was regarded as possibly a simple transformation by the metal catalyst of a symmetry-forbidden reaction into a symmetry-allowed reaction. Mango and Schachtschneider[27] advanced this idea to explain metal–catalyzed reactions in which a cyclobutane ring is converted into two olefin groups[28]:

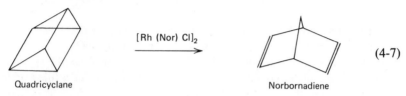

$$[Rh (Nor) Cl]_2 \tag{4-7}$$

Quadricyclane   Norbornadiene

Regardless of whether the proposed mechanism applies to this reaction (the metal catalyst may simply serve to stabilize an intermediate after rupture of a single CC bond), it is illuminating because it demonstrates the radical transformation that a metal atom can bring to an orbital symmetry correlation diagram.

Consider then the two face-on olefins, with a d^8 metal atom lying in P_{inter} (Fig. 4-16), slightly in front of the plane of the four carbon atoms. We are particularly concerned with those metal orbitals that may have the appropriate symmetry to mix or interact with $\pi - \pi'$ *(SA)* (Fig. 4-3), the crucial orbital that rises in energy when the molecules come into contact. Of the five metal d orbitals, only d_{xz} and d_{xy} are antisymmetric with respect to P_{inter}; the latter, however, is also antisymmetric with respect to P_{intra}. Hence only d_{xz} has the appropriate *SA* symmetry. Furthermore, through its back lobes, it points right at the two CC bonds. From ligand-field theory, we expect it to be the highest in energy of the metal d orbitals. If the metal has a d^8 configuration, d_{xz} *(SA)* will be empty. The mixing between d_{xz} *(SA)* and $\pi - \pi'$ *(SA)* now leads to an empty *(SA)*$_2$ orbital—mainly on the metal—and a filled *(SA)*$_1$ orbital, mainly on the olefins:

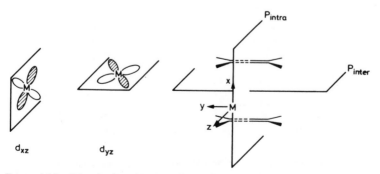

FIGURE 4-16. Dimerization of two ethylene molecules in the presence of a metal atom.

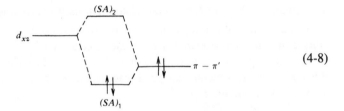

$$(4\text{-}8)$$

Thus an empty low-lying orbital of *SA* symmetry is provided that can correlate directly with the high-lying *SA* orbital ($\sigma^* + \sigma'^*$) of product.

By the same token, the filled d_{yz} orbital of metal that has *AS* symmetry, by mixing with the olefin combination $\pi^* + \pi'^*$ *(AS)*, provides a filled orbital that can correlate with the low-lying *AS* orbital ($\sigma - \sigma'$) of product. The final orbital correlations are shown in Fig. 4-17. On the product side, the two metal orbitals d_{xz} and d_{yz} that are considered to be pure again, are very close in energy since each points at two of the four cyclobutane CC bonds. Between them, they contain one pair of electrons.

All told, the activation energy of the reaction has been considerably lowered by the metal atom: the electron pair in $\pi + \pi'$ that had formally to rise to antibonding levels now goes only to the nonbonding metal level, and even that increase is compensated by the stabilization of a metal pair that drops to bonding ($\sigma - \sigma'$) level. We can think of the metal as catalyzing the reaction by receiving the awkward *SA* pair of electrons in its empty d_{xz} orbital and redonating it to the "substrate" olefins through its d_{yz} orbital. The metal atom functions as a "symmetry switch." Alternatively, the great resemblance of Figs. 4-17 and 4-5 suggests that the *additional pair* of electrons brought in by the metal, by yielding six electrons instead of four along the reaction path, in some manner gives "aromatic character" to the

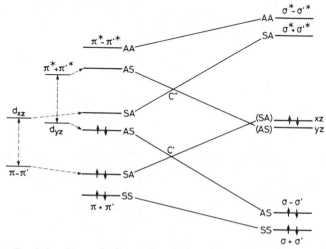

FIGURE 4-17.   Correlation diagram for face-on dimerization of two ethylene molecules in the presence of a metal atom. Symmetries respectively relative to P_{intra} and P_{inter}. Compare with Fig. 4-3.

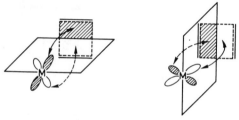

FIGURE 4-18.   Favorable interaction between metal orbitals and the two nonbonding orbitals of the antiaromatic cyclobutadienoid transition state.

transition state. This aromatic character does exist and is a consequence of the changes in both number of electrons and in orbital characteristics. Consider *both* metal orbitals in addition to the four basis carbon orbitals (Fig. 4-3). As demonstrated long ago by Longuet-Higgins and Orgel,[29] these metal orbitals have the appropriate symmetry to interact favorably with the two nonbonding orbitals (Fig. 3-7) of the antiaromatic cyclobutadienoid transition state (Fig. 4-18). Hence the two nonbonding orbitals (point C, Fig. 4-3) are transformed into a bonding pair (point C', Fig. 4-17) and an antibonding pair (point C''). The bonding pair can accommodate four electrons that, with the two low-lying olefinic SS electrons, give a total of six bonding electrons, the number required for aromaticity.

4.8   Effect of Electronic Excitation on Orbital-Symmetry-Controlled Reactions

Thus far we have considered orbital symmetry control of thermal reactions, for molecules in their ground electronic states. How is the correlation diagram affected if a molecule is excited? Woodward and Hoffmann limited their considerations to singly excited molecules, for instance, the dimerization of a singly excited ethylene molecule with a ground ethylene molecule. The excited electron is placed in the lowest molecular orbital of the pair (Fig. 4-19)—a description that, although different from that of the very appropriate exciton model, does account for the fact that the excitation is delocalized between the two weakly interacting molecules. The hole, concomitantly, is in the top bonding combination. The correlation diagram now leads from excited reactant to a $\sigma \rightarrow \sigma^*$ excited cyclobutane molecule and "there is no symmetry-imposed barrier to this transformation."[4] Aware of the troublesome fact that the $\sigma\sigma^*$ product state should be somewhat higher than the $\pi\pi^*$ reactant state, Woodward and Hoffmann point out that "the fact that the product state which correlates directly with reactant may lie higher in energy than the latter constitutes no special problem in orbital symmetry terms—though admittedly there is still much to be learned about the detailed physical nature of the processes accompanying the necessary energy cascade from electronically excited to ground states in such instances."

Yet this point is in all likelihood the single most important weakness in the theory of orbital symmetry control: the theory is insufficiently discriminate to specify which excited state is effectively responsible for the photochemical reaction. Although the predicted reversal of behavior—from thermally difficult to photo-

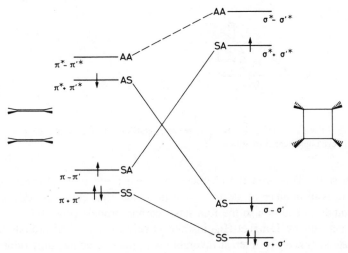

FIGURE 4-19.   Correlation diagram for face-on dimerization of ethylene with a singly excited partner (see Fig. 4-3).

chemically facile reaction—is observed experimentally,[30] it would have been rational to expect the appropriate correlation diagram to correlate an *excited reactant* with a *ground product*. Such correlations exist for many photochemical reactions (see Section 5.5). In the present case one may wonder if the effective state is not a *doubly excited* ethylene pair, obtained by interaction of a $(\pi, \pi^*)^2$ molecule with a ground molecule. Figure 4-20 shows that the doubly excited reactant pair does indeed correlate with ground product. In the mid-1960s the possibility of reactivity of a state represented by a doubly excited configuration seemed remote. Yet we see in Section 5.3 that in olefin photodimerization, as in other reactions that are symmetry forbidden in the ground state, a doubly excited state *is* indeed responsible for the major part of the observed reaction pathway.

Returning now to the Diels-Alder reaction (Fig. 4-5), we note that excitation of an electron has an effect opposite to that for ethylene dimerization. An electron is lifted from the strongly descending *A* orbital to a steeply rising *S* orbital, and

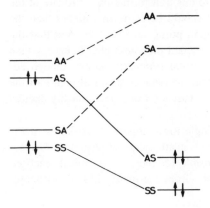

FIGURE 4-20.   The same correlation diagram for a doubly excited ethylene pair.

symmetry forbiddenness is brought in (higher excited states, like the lowest $S \rightarrow A$ configuration, might remain allowed). It is this alternation of allowedness and forbiddenness—from four to six electrons, or from thermal to photochemical—that underscores the brilliance of Woodward and Hoffmann's discovery.

4.9 Symmetry Control Without Symmetry

The experimental verification of the Woodward-Hoffmann predictions goes far beyond the unsubstituted model molecules for which the correlation diagrams were first drawn out. Hence the two additions

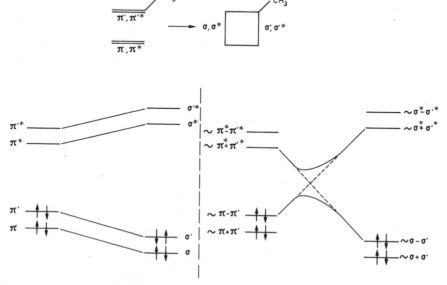

are expected to obey the same orbital-symmetry restrictions. Yet in the case of propylene cycloaddition to ethylene, both planes of symmetry P_{inter} and P_{intra} (Fig. 4-2) have disappeared! In the absence of any controlling symmetry element a brute force correlation diagram, based on the fact that all orbitals have the same symmetry in the point group of the pair, gives the impression that the reaction is allowed[4] (Fig. 4-21).

FIGURE 4-21.   Correlation diagram for addition of ethylene to propylene[4] (sign $\sim$ indicates that the combinations do not have equal weights on ethylene and propylene side): *(left)* incorrect (all orbitals have same symmetry); *(right)* correct.

What is really happening is that we are faced with what might be called a case of *pseudosymmetry*. Consideration of the molecular pair shows that there is *almost* a plane P_{intra} of symmetry cutting through the CC bonds and almost a plane P_{inter} of symmetry parallel to both molecular planes. Hence the orbitals of the interacting ethylene-propylene pair combine strongly as do those for the ethylene dimer pair, with the *reservation* that the crossing of levels $\pi - \pi'$ *(SA)* and $\pi^* + \pi'^*$ *(AS)* in Fig. 4-3 is now "avoided" because of the noncrossing rule for orbitals of same symmetry.[16]

Let us consider this pseudosymmetry effect in more detail. For identical monomers, the orbitals have different symmetry and their crossing is allowed. Mathematically, the matrix element of the one-electron Hamiltonian between the two orbitals [see (4-6)] vanishes:

$$\langle \pi - \pi' | H_{eff} | \pi^* + \pi'^* \rangle = 0 \tag{4-10}$$

since clearly the local elements

$$\langle \pi | H_{eff} | \pi^* \rangle = \langle \pi' | H_{eff} | \pi'^* \rangle \tag{4-11}$$

are the same on both molecules, and the cross-terms

$$\langle \pi | H_{eff} | \pi'^* \rangle = \langle \pi' | H_{eff} | \pi^* \rangle \tag{4-12}$$

are also identical (H_{eff} is the effective Hamiltonian for the full "supersystem"). When one hydrogen atom is substituted by a methyl group, (4-10) breaks down for two reasons:

1. The combinations of the two double-bond orbitals are no longer quite equally weighted on both sides: the methyl group pushes the propylene orbitals π' and π'^* up in energy relative to the ethylene orbitals π and π^* and the proper intermolecular combinations become

$$\begin{cases} (1 + \epsilon)\pi + (1 - \epsilon)\pi' \\ (1 - \epsilon)\pi - (1 + \epsilon)\pi' \end{cases} \text{(bonding)}$$

and $\tag{4-13}$

$$\begin{cases} (1 + \epsilon)\pi^* + (1 - \epsilon)\pi'^* \\ (1 - \epsilon)\pi^* - (1 + \epsilon)\pi'^* \end{cases} \text{(antibonding)}$$

where ϵ is a measure of the Coulombic perturbation caused by the methyl substitution. Hence the matrix element (4-10) has a nonvanishing term, to order ϵ:

$$-2\epsilon(\langle \pi | H_{eff} | \pi'^* \rangle + \langle \pi' | H_{eff} | \pi^* \rangle) \neq 0 \tag{4-14}$$

which is not compensated for, even if (4-11) and (4-12) were still to hold.

2. Equations (4-11) and (4-12) do not hold rigorously any more. In (4-11) the effect of methyl group potential field on $\pi\pi^*$ in the first term is different from the effect of hydrogen atom potential field acting on $\pi'\pi'^*$ in the second term. Hence there are small nonvanishing contributions

$$\langle\pi|H_{\text{eff}}|\pi^*\rangle - \langle\pi'|H_{\text{eff}}|\pi'^*\rangle \neq 0, \qquad \langle\pi|H_{\text{eff}}|\pi'^*\rangle - \langle\pi'|H_{\text{eff}}|\pi^*\rangle \neq 0 \quad (4\text{-}15)$$

which are likely to be of order ϵ too.

To draw out the orbital correlation diagram, the correct procedure requires first drawing in dotted lines the *intended* correlations identical with those of Fig. 4-3, and then accounting for the avoided crossing by joining the dotted line originating from $\pi - \pi'$ and descending dotted line terminating at $\sigma - \sigma'$ in a curved manner, giving the solid line shown in Fig. 4-21, RHS. A second curved line is obtained for the antibonding levels. The resultant barrier reflects the avoided crossing and the inherent forbiddenness of the reaction. Its height is difficult to estimate without an actual numerical calculation, but as long as the substitutional perturbation is small, the crossing should be nearly effective and the barrier high. Pearson[1] points out that the number and kind of nodal surfaces for the wave functions have not changed, even though these surfaces may be distorted and deformed. Forbiddenness remains because the wave function maintains its "topological identity".[31]

If the dissymmetry increases, clearly the argument breaks down. In the extreme limit of donor-acceptor substitution of each double bond, we could have

$$(4\text{-}16)$$

that must clearly become facile cycloadditions because of the favorable ionic interactions. This was first pointed out by Epiotis.[32,33] In the left-hand system the local atomic orbitals on each double bond have such different energies that it is more appropriate to start by combining the two low-energy atomic orbitals (on the different molecules) adjacent to acceptors and then do the same for the high-energy atomic orbitals adjacent to donors. Use of the pseudo-C_2 axis of the system as discriminating symmetry element leads formally to an allowed reaction.*

In the right-hand system the intramolecular symmetry plane P_{intra} is still present, or nearly so. Hence a correlation diagram can be drawn out in the normal fashion,[32] with the novel feature that the starting orbitals—π on electron-rich olefin, π' on

---

*The C_2 axis here is not a proper symmetry element as defined by Woodward and Hoffmann for studying pericyclic reactions[4]—one that should cut through bonds being made or being broken. But this reaction, precisely, falls out of the family of normal pericyclic reactions.

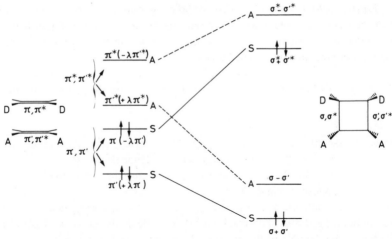

FIGURE 4-22.   Orbital correlation diagram for addition of electron-rich olefin to electron-poor olefin.[32]

electron-poor olefin—are well separated in energy. Hence, contrary to the case of two ethylenes, or even to ethylene plus propylene, they hardly combine at all to form intermolecular combinations. The highest occupied orbital, essentially π on electron-rich olefin, lies even close to the lowest unoccupied orbital, essentially π'^* on electron-poor olefin. Although the correlations (Fig. 4-22) indicate a formally forbidden reaction with an orbital crossing, as ground-state reactant correlates with doubly excited product, the crossing occurs so early along the pathway that the one-electron barrier should be small. Two-electron effects introduced by configuration interaction (to give the proper states, Section 5.1) can then wipe it out entirely. The reaction becomes allowed for ''ionic'' reasons. An important *caveat* in the analysis of such reactions is that they may actually occur through an entirely different mechanism, involving an initial *electron*-transfer step (Section 8.9).[34]

4.10 Orbital Correlations in "Maximum Symmetry"

Halevi and other authors have pointed out[35–38] that there is some loss of information in the conventional Woodward-Hoffmann correlation diagrams which use only one or two controlling symmetry elements. By using the full symmetry common to reactant and product, the allowed or forbidden character comes forth naturally and in addition it is possible to find motions that transform a forbidden reaction to an allowed one.

The simplest analysis is again that of the face-on dimerization of ethylene, in which both reactants and product—with the two new CC bonds assumed still slightly stretched relative to those already present—belong to the D_{2h} point group. A ''correspondence diagram''[35] is constructed by ordering reactant and product orbitals energetically and labeling them according to their full symmetry species in the D_{2h} group (Fig. 4-23). Except for the labels, the diagram is identical with that of Fig. 4-3. But we now have the possibility of determining the nature of the

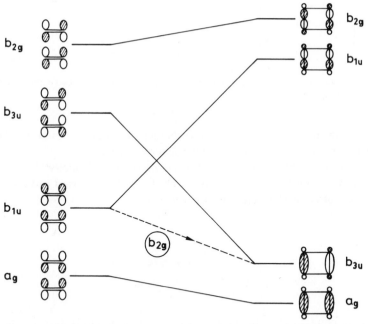

FIGURE 4-23. Correspondence diagram for face-on dimerization of ethylene.[35] A common D_{2h} symmetry is assumed. Reprinted with permission from E. A. Halevi, *Helv. Chim. Acta* **58**, 2136 (1975). Copyright 1975 Helvetica Chimica Acta.

displacement—specifically, the symmetry of the appropriate motion—which would render the reaction allowed.

The normal face-on approach can be considered to have the full (a_g) symmetry of the D_{2h} point group common to reactant and product. The inducing symmetry coordinate that would make the reaction allowed must correlate $\pi - \pi'$ (b_{1u}) with $\sigma - \sigma'$ (b_{3u}); it must, therefore, transform as the direct product

$$b_{1u} \times b_{3u} = b_{2g} \qquad (4\text{-}17)$$

The b_{2g} motion in a rectangle of carbon atoms, if it involves solely CC stretching, tends to shorten one diagonal of the rectangle while it lengthens the other (Fig. 4-24). Hence the correspondence diagram indicates that the approaching ethylene molecules will want to form a transoid tetramethylene diradical:

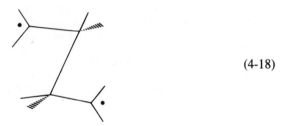

$$(4\text{-}18)$$

FIGURE 4-24.   The b_{2g} motion for the coplanar approach of two ethylene molecules (D_{2h} symmetry).

[compare with (4-3)]. Such an extended diradical does appear to be involved in the reaction according to quantum-mechanical calculations.[39]

The "orbital correspondence analysis in maximum symmetry"[35] sometimes offers, however, predictions at variance with those of Woodward and Hoffmann. The isomerization of benzvalene to benzene

$$\text{(benzvalene)} \longrightarrow \text{(benzene)} \tag{4-19}$$

is a case in point. Use of the full C_{2v} symmetry common to both molecules leads to an "allowed" prediction at variance with experiment, whereas consideration of the local symmetry[40] indicates a forbidden reaction. Nevertheless, the method is important because it gives a foretaste of other methods where orbital symmetry of reactants determines the symmetry of facile reaction modes (Chapter 6).

References

1. R. G. Pearson, *Symmetry Rules for Chemical Reactions* (Wiley-Interscience, New York, 1976); *Orbital Symmetry Papers* (American Chemical Society Reprint Collection, Washington, 1974); N. T. Anh, *Les Règles de Woodward-Hoffmann* (Ediscience, Paris, 1970).
2. E. Wigner and E. E. Witmer, Z. *Physik* **51**, 859 (1928).
3. R. S. Mulliken, *Rev. Mod. Phys.* **4**, 1 (1932).
4. R. B. Woodward and R. Hoffmann, *The Conservation of Orbital Symmetry* (Verlag Chemie, Weinheim, 1970); R. B. Woodward and R. Hoffmann, *Angew. Chem. Internatl. Ed.* **8**, 781 (1969).
5. R. B. Woodward and R. Hoffmann, *J. Am. Chem. Soc.* **87**, 395 (1965).
6. R. Hoffmann and R. B. Woodward, *J. Am. Chem. Soc.* **87**, 2046 (1965).
7. R. Srinivasan and J. N. C. Hsu, *Chem. Commun.* 1213 (1972); R. Huisgen, *Acc. Chem. Res.* **10**, 199 (1977).
8. M. J. S. Dewar, *Tetrahedron Suppl.* **8**, 75 (1966); *Angew. Chem. Internatl. Ed.* **10**, 761 (1971).
9. H. E. Zimmerman, *J. Am. Chem. Soc.* **88**, 1564 (1966); *Acc. Chem. Res.* **4**, 272 (1971).
10. M. G. Evans and E. Warhurst, *Transact. Faraday Soc.* **34**, 614 (1938); M. G. Evans, ibid. **35**, 824 (1939).
11. A. Streitwieser, *Molecular Orbital Theory for Organic Chemists* (Wiley, New York, 1961); E. Heilbronner and H. Bock, *Das HMO-Modell und seine Anwendung; Grundlagen und Handhabung* (Verlag-Chemie, Weinstein, 1968); L. Salem, *The Molecular Orbital Theory of Conjugated Systems* (Benjamin, New York, 1966).

12. M. Uchiyama, T. Tomioka, and A. Amano, *J. Phys. Chem.* **68**, 1878 (1964); W. Tsang, *J. Chem. Phys.* **42**, 1805 (1965).

13. K. N. Houk, R. W. Gandour, R. W. Strozier, N. G. Rondan, and L. A. Paquette, *J. Am. Chem. Soc.* **101**, 6797 (1979).

14. Ref. 4, pp. 25–26.

15. P. Millie, *Bull. Soc. Chim. Fr.* 4031 (1966); H. Nohira, *Tetrahedron Lett.* 2573 (1974).

16. J. von Neumann and E. Wigner, *Physik. Z.* **30**, 467 (1929); C. A. Coulson, *Valence,* 2nd ed. (Oxford University Press, Oxford, 1961), p. 65.

17. H. Hosoya, *Kagaku no Ryoiki* **28**, 45 (1974).

18. R. E. Townshend, G. Ramunni, G. Segal, W. J. Hehre, and L. Salem, *J. Am. Chem. Soc.* **98**, 2190 (1976); W. L. Jorgensen, private communication to the author (1972).

19. A. Devaquet, A. Sevin, and B. Bigot, *J. Am. Chem. Soc.* **100**, 2009 (1978); see also ref. 22.

20. E. B. Wilson, *J. Chem. Phys.* **63**, 4870 (1975); W. T. Dixon, *J. Chem. Soc., Faraday Transact. II,* **74**, 511 (1978); K. F. Herzfeld, *Z. Naturforsch.* **3a**, 457 (1948); *Rev. Mod. Phys.* **41**, 527 (1949).

21. W. A. Goddard, *J. Am. Chem. Soc.* **94**, 793 (1972).

22. C. Trindle and O. Sinanoglu, *J. Am. Chem. Soc.* **91**, 4054 (1969).

23. W. J. van der Hart, J. J. C. Mulder, and L. J. Oosterhoff, *J. Am. Chem. Soc.* **94**, 5724 (1972); D. M. Silver and M. Karplus, ibid. **97**, 2645 (1975).

24. E. Heilbronner, *Tetrahedron Lett.* 1923 (1964).

25. L. Salem, *Nouv. J. Chimie* **2**, 559 (1978).

26. J. Mathieu, *Comptes Rendus Acad. Sci. (Paris)* **274**, 81 (1972); A. Rassat, *Comptes Rendus Acad. Sci. (Paris),* **274**, 730 (1972).

27. F. D. Mango and J. Schachtschneider, *J. Am. Chem. Soc.* **89**, 2484 (1967).

28. H. Hogeveen and H. C. Volger, *J. Am. Chem. Soc.* **89**, 2486 (1967).

29. H. C. Longuet-Higgins and L. E. Orgel, *J. Chem. Soc.* 1969 (1956).

30. W. L. Dilling, *Chem. Res.* **66**, 373 (1966).

31. C. Trindle, *J. Am. Chem. Soc.* **92**, 3251 (1970).

32. N. D. Epiotis, *J. Am. Chem. Soc.* **95**, 1191 (1973).

33. N. D. Epiotis, *J. Am. Chem. Soc.* **94**, 1924 (1972).

34. N. Kornblum, *Angew. Chem. Internatl. Ed.* **14**, 734 (1975).

35. E. A. Halevi, *Helv. Chim. Acta.* **58**, 2136 (1975); J. Katriel and E. A. Halevi, *Theoret. Chim. Acta (Berl.)* **40**, 1 (1975); E. A. Halevi, *Nouv. J. Chim.* **1**, 229 (1977).

36. H. Metiu, J. Ross, and T. F. George, *Chem. Phys.* **11**, 259 (1975).

37. Y. N. Chiu, *J. Chem. Phys.* **64**, 2997 (1976).

38. R. Wallace, *Chem. Phys.* **37**, 285 (1979).

39. R. Hoffmann, S. Swaminathan, B. G. Odell, and R. Gleiter, *J. Am. Chem. Soc.* **92**, 7091 (1970); G. A. Segal, *J. Am. Chem. Soc.* **96**, 7892 (1974).

40. N. T. Anh, editor's note in E. A. Halevi, *Nouv. J. Chim.* **1**, 229 (1977).

5

Electronic Control by State Symmetry

The power of orbital symmetry control, however great, is limited by the ultimate control, that of electronic states. The preservation of the symmetry of an orbital implies the conservation of some form of momentum[1]: linear momentum in the case of symmetry with respect to a plane and angular momentum in the case of symmetry around an axis. Hence orbital symmetry preservation is equivalent to the conservation of *individual* electronic orbital angular momenta. The preservation of state symmetry, on the other hand, implies the conservation of *total* electronic orbital angular momentum.[1] This last principle derives from the Wigner-Witmer rules,[2–4] which state that reactions are "feasible" or "unfeasible" depending on whether total state symmetry is conserved. In this chapter we encounter cases where state symmetry gives little information that is not already provided at the orbital level, but also other examples where state symmetry exerts a control that was nonexistent at the orbital level.

5.1 From Orbital Symmetry Correlations to State Symmetry Correlations

Here and in the following sections we leave the orbital picture, in which the electrons are considered to move independently, each in their own spatial function, for the state picture, in which all the electrons are mutually and instantaneously interacting. The correspondence between the two descriptions depends on whether the state is well represented by a single distribution of the electrons among the different orbitals ("configuration") or whether it requires several such distributions ("configurations"):

For *(a)* a closed-shell ground state, with no low-lying empty orbitals, or *(b)* a singly-excited state with two half-empty orbitals well separated from the remaining manifold, a single configuration suffices*:

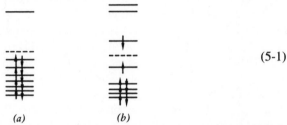

$$(5\text{-}1)$$

(a)   *(b)*

*In the excited-state singlet, or $S_z = 0$ triplet, two *determinants* are needed, but both represent the same electron distribution.

For an open-shell singly excited state in which one of the two half-empty orbitals lies close to another orbital, two configurations are necessary, as shown in the two different examples in (5-2):

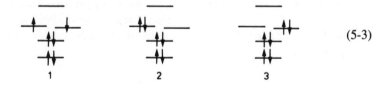

$$(5\text{-}2)$$

Other cases of nearly equienergetic configurations arise whenever, for a given orbital pattern, two different one-electron excitations have the same energy. The proper wave function for the state is a combination of the two configurations, assumed to have the same spatial symmetry.

For an open-shell diradical–like system in which two nearly equienergetic orbitals compete for two electrons, the singlet states are generally an appropriate combination of three configurations [see (3-9)]:

$$(5\text{-}3)$$

The operator that leads to the correct states by mixing the "one-electron" configurations together is the potential energy for electron repulsion, taken fully and not in some average manner. The corresponding term in the Hamiltonian is

$$H' = \sum_{i<j} \frac{1}{r_{ij}} - \overline{\sum_{i<j} \frac{1}{r_{ij}}} \tag{5-4}$$

since this term is what separates an approximate independent–electron description from the correct description in which all electrons are interacting instantaneously.*

Let us consider then the correlation of states for the helium-helium interaction (Section 4.1). Using the orbital correlations in Fig. 4-1, we immediately obtain the correlations between configurations:

*Even in the Hartree-Fock approximation (Section 1.5) electron repulsion is included only in an average manner. Sinanoglu[5] has shown that the correlation energy correcting the Hartree-Fock energy is obtained by the perturbation Hamiltonian (5-4).

$$(1s\sigma_g)^2 \, (1s\sigma_u)^2 \rightarrow (1s_{Be})^2 \, (2p_{Be})^2$$

$$(1s\sigma_g)^2 \, (1s\sigma_u) \, (2s\sigma_g) \rightarrow (1s_{Be})^2 \, (2s_{Be}) \, (2p_{Be}) \qquad (5\text{-}5)$$

$$(1s\sigma_g)^2 \, (2s\sigma_g)^2 \rightarrow (1s_{Be})^2 \, (2s_{Be})^2$$

The energetic behavior of these three configurations is shown in Fig. 5-1. The singly excited configuration has Σ_u symmetry and, as in (5-1), describes appropriately the lowest excited *state* of u symmetry for both reactant and product. The other two configurations, whose energies follow the dashed lines, have Σ_g symmetry and intersect. As it is a direct consequence of the orbital intersection point C in Fig. 4-1, this intersection of configurations with identical symmetry and equal energy is also a flashing indicator that mixing of configurations is required to describe the correct states.

The mixing of the two configurations at and around the crossing point yields two states, whose form and energy are given by the roots of the secular determinant:

$$\begin{vmatrix} E_1 - E & H'_{12} \\ H'_{12} & E_2 - E \end{vmatrix} = 0 \qquad (5\text{-}6)$$

where

$$\left. \begin{array}{l} E_1 = \text{energy of } (1s\sigma_g)^2 \, (1s\sigma_u)^2 \\[1em] E_2 = \text{energy of } (1s\sigma_g)^2 \, (2s\sigma_g)^2 \end{array} \right\} \quad \begin{array}{l} \text{calculated for the effective} \\ \text{Hamiltonian (Hartree-Fock or other)} \\ \text{used to obtain the orbital energies} \end{array} \qquad (5\text{-}7)$$

and

$$H'_{12} = \langle |1s\sigma_g \, \overline{1s\sigma_g} \, 1s\sigma_u \, \overline{1s\sigma_u}| \, |H'| \, |1s\sigma_g \, \overline{1s\sigma_g} \, 2s\sigma_g \, \overline{2s\sigma_g}| \rangle = K_{12} \qquad (5\text{-}8)$$

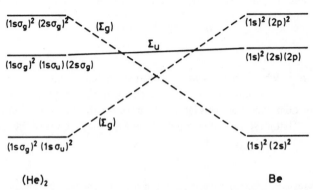

FIGURE 5-1.   Correlations between *configurations* of $(\text{He})_2$ and beryllium.

The matrix element K_{12} is simply the exchange integral

$$K_{12} = \int\int 1s\sigma_u(1)\, 2s\sigma_g(1)\, \frac{1}{r_{12}}\, 1s\sigma_u(2)\, 2s\sigma_g(2)\, d\tau_1\, d\tau_2 > 0 \qquad (5\text{-}9)$$

This integral may be relatively large since the exchange density $1s\sigma_u\, 2s\sigma_g$ concerns two orbitals of different inversion symmetry, but concentrated in the same region of space. The roots of (5-6) are given by

$$E = \frac{1}{2}\{E_1 + E_2 \pm \sqrt{(E_1 - E_2)^2 + 4K_{12}^2}\} \qquad (5\text{-}10)$$

At the crossing point ($E_1 = E_2$) the energy *shift* relative to the one–electron level $(E_1 + E_2)/2$ is directly proportional to the exchange integral

$$\Delta E = \pm K_{12} \qquad (5\text{-}11)$$

As one departs from the crossing point, this shift becomes smaller and smaller and is finally given by the perturbation expression

$$\Delta E \quad \text{for} \quad E_1 = -\frac{K_{12}^2}{E_2 - E_1}$$

$$(E_1 < E_2) \qquad\qquad (5\text{-}12)$$

$$\Delta E \quad \text{for} \quad E_2 = +\frac{K_{12}^2}{E_2 - E_1}$$

The final state correlations are illustrated in Fig. 5-2. Instead of the "intended" intersecting straight lines (dashed), smooth curves (full lines) are obtained for both ground electronic state and doubly excited state of symmetry $^1\Sigma_g$. The curves tend toward each other and then bend away, as the mixing between configurations repels the two states. They behave as if their crossing (real at the configuration level)

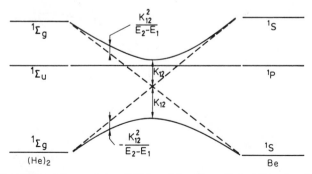

FIGURE 5-2. Correlations between *states* of $(He)_2$ and beryllium. Dashed lines show intended correlations due to configurations.

were "avoided." The ground state of reactant now correlates with ground state of product; the reaction is "feasible".[1] But it stays orbitally forbidden since the downshift K_{12} at the barrier is not large enough to offset the large rise in energy due to the double occupancy of the antibonding $1s\sigma_u$ orbital.

5.2 State Correlations for Woodward-Hoffmann (Pericyclic) Reactions: Thermal Aspects

Longuet-Higgins and Abrahamson, in one of the earliest orbital symmetry papers,[6] demonstrated how orbital symmetry correlations should be translated into state symmetry correlations. The principle of the method is the same as that expounded in Section 5.1. Orbital correlations are used to draw the correct correlations for the *configurations*, which in turn give *state* correlations by appropriate "avoided" crossings between those configurations that have common symmetry and equal energies. The result for the face-on dimerization of ethylene is shown in Fig. 5-3. (This figure could be refined by placing the ground state of product at its appropriate energy some 75 kJ/mole below reactant.) Again there is an intersection between ground and doubly excited configuration that disappears when electron repulsion is included. However, the "intended" crossing remains in the form of the large barrier on the ground surface.

A crucial question concerns the size of the exchange integral

$$K = \iint SA(1) \cdot AS(1) \frac{1}{r_{12}} SA(2) \, AS(2) \, d\tau_1 \, d\tau_2 \tag{5-13}$$

where *SA* and *AS* are the intersecting molecular orbitals shown in Fig. 4-3. The exchange density $SA \times AS$ is sketched in Fig. 5-4. It has the form

$$SA \times AS \approx a^2 - b^2 - c^2 + d^2 \tag{5-14}$$

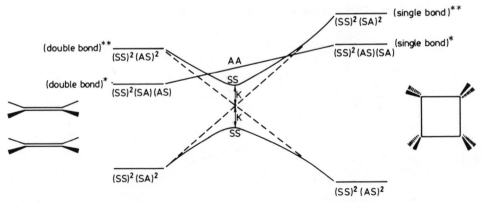

FIGURE 5-3.   State correlations for face-on dimerization of ethylene. Dashed lines indicate correlation of configurations, which are given for reactants and product. State symmetries with respect to the two symmetry planes are given in the middle (see Figs. 4-2 and 4-3). Double stars indicate double excitation.

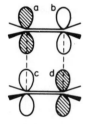

FIGURE 5-4.   Exchange density $SA \times AS$ for the ethylene dimer orbitals.

where a, b, c, and d are the four atomic π orbitals. The zero-differential–overlap approximation (Section 1.9) leads directly to

$$K \approx 4\gamma_{11} - 8\gamma_{12} + 4\gamma_{13} \qquad (5\text{-}15)$$
$$= 4(\gamma_{11} - \gamma_{12}) - 4(\gamma_{12} - \gamma_{13})$$

where γ_{11} is a self-repulsion integral, γ_{12} is a repulsion between electrons on adjacent orbitals in the rectangle (γ_{ab} or γ_{ac}), and γ_{13} is a diagonal repulsion. Since the integral γ decreases slowly with distance, K cannot be too large. Ab initio calculations[7] give

$$K = 2.5 \text{ eV} \qquad (5\text{-}16)$$

This is significant, but still too small to erase the barrier imposed by the orbital correlation. The latter can be estimated as half* the energy required to promote two π electrons into a σ^* orbital, or a π,σ^* promotion energy. This should lie halfway between a π,π^* (6-7 eV) and a σ,σ^* (10 eV) energy—at about 8eV. There remains a large (5.5 eV or 535 kJ/mole) activation energy for the forward reaction that is sufficient to forbid any "concerted" (Sections 2.1 and 4.2) dimerization. The reverse reaction has an even larger calculated activation energy since cyclobutane lies lower in energy than two ethylene molecules. The experimental results[8] show an activation energy of about 260 kJ/mole for the thermal (400°) decomposition of cyclobutane. This proves convincingly that the reaction does not proceed in a concerted manner along a rectangular decomposition path.

Let us now construct the state correlation diagram for the Diels-Alder reaction. Using Fig. 4-5, we obtain the reactant and product configurations shown in Fig. 5-5. (This figure could be refined by placing ground state of product at its appropriate energy some 135 kJ/mole below reactant.) The ground configuration of reactant ($S^2S^2A^2$) correlates directly with ground configuration of product ($S^2A^2S^2$). Since the respective states are well represented by single configurations, the ground states of reactant and product are linked smoothly together. The lowest excited configuration of reactants corresponds to π,π^* excitation on the butadiene fraction (S^2S^2AS). This configuration correlates with a high-energy σ,σ^* ($S^2AS^2$$_S$) configuration of cyclohexene, in accord with the orbital-symmetry forbiddenness of the

---

*The orbital intersection that locates the transition state occurs about halfway along the pathway.

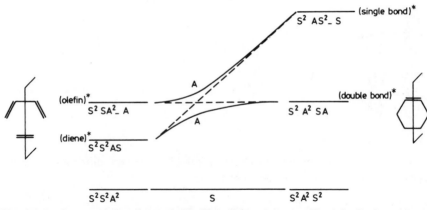

FIGURE 5-5.   State correlation diagram for the Diels-Alder reaction (see Fig. 4-5). Configurations are given for reactant and product. A dash indicates that the corresponding orbital (in order of increasing energy) is vacant. State symmetries are written in the middle.

reaction. On the other hand, the higher-energy π,π^* configuration for excitation on ethylene correlates with the olefinic excited configuration of similar energy in product. Hence the configurations, both of symmetry A, intersect; the crossing becomes "avoided" at the level of the states.

5.3   State Correlations for Woodward-Hoffmann (Pericyclic) Reactions: Photochemical Aspects

We saw in Section 4.8 that an orbital correlation diagram proves the photochemical dimerization of olefins to be orbital-symmetry allowed but is not too explicit as to the state responsible for the photochemical reaction. In their original state correlation diagram, Longuet-Higgins and Abrahamson kept the excited SS and AA states well separated, with no indication of the intersection that might exist between them (Fig. 5-3). Yet such an intersection is plausible if the descending nature of the first state (that tends toward ground product) and the rising nature of the second state (lower energy for π,π^* excitation in reactant relative to σ,σ^* excitation in product) are accounted for. Hence not only is the doubly excited state of reactant the one that truly "descends," but it also becomes the *lowest* excited state along the reaction pathway.

The notion that a doubly-excited symmetric excited state should be the driving force for a photoinduced symmetry-allowed reaction was first put forward by van der Lugt and Oosterhoff.[9] Studying the disrotatory ring closure of butadiene to cyclobutene (Section 6.3) in a four-electron valence-bond model, they found potential surfaces for ground, singly excited and "doubly excited" states* similar to

---

*The authors considered the lowest antisymmetric excited state of butadiene (major configuration: $\pi_2 \rightarrow \pi_3^*$) and the lowest symmetric excited state (with a mixture of configurations $\pi_1 \rightarrow \pi_3^* - \pi_2 \rightarrow \pi_4^*$ and $\pi_2^2 \rightarrow \pi_3^{*2}$) relative to the plane of symmetry reserved along the reaction pathway. For the electronic states of vertically excited butadiene, see Ref. 10.

the curves in Fig. 5-3. They assumed that the excited reactant starts off on the singly excited state surface and switches over to the "doubly excited" state by a radiationless transition near their intersection (Fig. 5-3). Greeted with some skepticism, this remarkable prediction—from a great scientist, Oosterhoff, who had already guessed at the role of orbital symmetry in chemical reactions[11]—was soon to be supported by further full ab initio calculations.[12] Figure 5-6 shows the ground and lowest excited-state potential-energy surfaces for the photochemical symmetry-allowed disrotatory cyclization of butadiene to cyclobutene. The most prominent feature is the sharp dip in the doubly excited state S_2, reminiscent of its "avoided crossing" with ground state (Section 5.2) and that brings it well below the singly excited state S_1. The latter, on the other hand, rises continuously from reactant to product, as does its triplet partner T_1 that runs parallel to S_1 at lower energies. The triplet actually finds a facile deactivation pathway through diene isomerization around one of the two double bonds (Fig. 3-5). Within the singlet manifold the most likely candidate for reaction is S_2, not only because it falls slightly lower than S_1, but especially because in doing so S_2 gets much closer to the ground state to which decay is ultimately necessary (Section 5.10). An early suggestion[13] that the spectroscopic singlet state of the diene decays to a "nonspectroscopic" excited singlet state is thus substantiated.

Recently similar results have come forth on the photochemical dimerization of ethylene[14]; again S_2 falls below S_1 in the region of its avoided crossing with S_0. In a sense the crucial role of the doubly excited state can be recognized as soon as one remembers (Table 3-12) that the lowest excited state of cyclobutadiene is the *symmetric* $^1A_{1g}$ state rather than the *antisymmetric* $^1B_{2g}$ state. As shown in Section 5.4, the former state, which is an "in-phase" zwitterionic state (Table 3-11), is

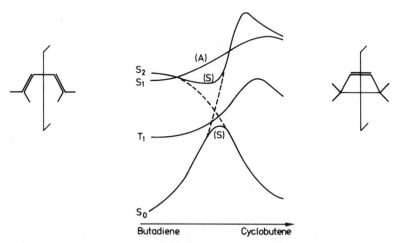

FIGURE 5-6. Potential-energy surfaces for the lowest states in the disrotatory closure of butadiene.[12] A linear internal coordinate pathway (Section 2.5) is chosen. Symmetries relative to vertical plane. Reprinted with permission from D. Grimbert, G. Segal, and A. Devaquet, *J. Am. Chem. Soc.* **97**, 6629 (1975). Copyright 1975 by American Chemical Society.

TABLE 5-1

Correlation Between Planar Butadiene and Cyclobutadienoid Antiaromatic Transition State[a]

	Butadiene	$\longrightarrow$	Cyclobutadienoid TS	
	State[10]	Configuration	Wave Function[b]	State
S_0	Ground (A_1)	$\pi_1^2\pi_2^2$	$\lvert\psi_l\bar{\psi}_m\rvert + \lvert\psi_m\bar{\psi}_l\rvert$	Singlet diradical $(^1B_{1g})$
T_1	Lowest triplet $(^3B_2)$	$^3(\pi_2 \rightarrow \pi_3^*)$	$\lvert\psi_l\psi_m\rvert$	Triplet diradical $(^3A_{2g})$
S_1	Singly excited valence singlet $(^1B_2)$	$^1(\pi_2 \rightarrow \pi_3^*)$	$\lvert\psi_l\bar{\psi}_l\rvert - \lvert\psi_m\bar{\psi}_m\rvert$	Out-of-phase zwitterionic $(^1B_{2g})$
S_2	"Doubly excited" valence singlet $(^1A_1)$	$\begin{cases}(\pi_2 \rightarrow \pi_3^*)^2\\\pi_1 \rightarrow \pi_3^* - \pi_2 \rightarrow\\\pi_4^*\end{cases}$	$\lvert\psi_l\bar{\psi}_l\rvert + \lvert\psi_m\bar{\psi}_m\rvert$	In-phase zwitterionic $(^1A_{1g})$

[a]Compare with Table 3-5. Slight differences with Table 3-5 disappear if rectangular orbitals are used for wave function. See Section 5.4 for a more detailed description of the $^1A_{1g}$ state of cyclobutadiene.
[b]As described by lozenged orbitals (see also Table 3-11).

stabilized by covalent terms that do not contribute to the "out-of-phase" latter state. Remembering that the transition states for ethylene dimerization or butadiene disrotatory closure are antiaromatic (Section 4.6) and have the electronic features of cyclobutadiene, we obtain directly the energy ordering due to Oosterhoff. The actual correlations between butadiene states and cyclobutadienoid states are shown in Table 5-1.

A final word concerns the orbital-symmetry-forbidden photochemical Diels-Alder reaction. Little is known about the excited surface of the reaction, but the rising nature of both singly and doubly excited surfaces indicated by the correlation diagram (Fig. 5-5) is likely to carry through in accurate calculations. Orbital forbiddenness should continue to control the reaction perfectly.

5.4 Nature of Excited Minima Lying Above a Cyclic Four-Atom Transition State

Michl has given an illuminating discussion[15] of the nature of the competing "singly excited" and "doubly excited" states that lie above the ground-state barrier in reaction diagrams such as Fig. 5-3 and 5-6. Choosing the H_4 system as model, he first correlates that states of square H_4 with those of two H_2 molecules at large separation (Fig. 5-7). These correlations should hold also for the cyclobutane $\rightarrow$ two ethylenes photodecomposition. For the in–plane rectangular approach: (1) ground-state S_0 (H_4) correlates with two ground-state H_2 molecules; (2) singly excited state S_1 (H_4) correlates with H_2 + singly excited ($^1\Sigma_u^+$) H_2^* (of which there are two possibilities, the other one leading necessarily to a higher state of square H_4); and (3) doubly excited state S_2 (H_4) correlates with two hydrogen molecules

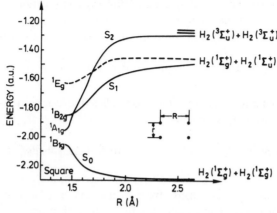

FIGURE 5-7. State correlation diagram for rectangular coplanar decomposition of H_4 to two H_2 molecules.[15] Dotted lines show higher state also obtained (in addition to S_1) from one ground H_2 plus one singly-excited H_2. Compare with Table 5-1 and Fig. 3-5. At $R = 2.75$ Å the H_2 lengths are the equilibrium values ($R_1 = R_2 = r = 0.76$ Å). As the molecules approach, the H_2 lengths increase, reaching $r = 1.43$ Å $= R$ at the squre. Adapted with permission from W. Gerhartz, R. D. Poshusta, and J. Michl, *J. Am. Chem. Soc.* **98**, 6427 (1976). Copyright 1976 by American Chemical Society.

both excited to their triplet state ($^3\Sigma_u^+$). Correlations 2 and 3 reveal the incomplete nature of the molecular orbital description of the ethylene dimer excited states in Section 5.2 and Fig. 5-3. Instead of a singly excited "configuration" or a doubly excited "configuration", these are really complex states, which, of course, could also be derived from the molecular orbital picture through appropriate configuration interaction and conversion from the orbitals of the dimer supersystem to those of the subsystems. The "double-triplet" nature of the lowest doubly excited state of the separate molecules is not too surprising if one remembers that (1) the second valence singlet of butadiene (S_2 in Table 5-1) can be conveniently described as two coupled triplets*

$$\uparrow \quad \uparrow \quad \diagdown \qquad \qquad (5\text{-}17)$$
$$\downarrow \quad \downarrow$$

and (2) one of the two singlets for four electrons in a square is obtained by coupling two two-center triplets (Table 1-1).

The bonding characteristics of S_1 and S_2 for the cyclobutadienoid square are most interesting. We saw (Section 3.9) that S_2 ($^1A_{1g}$ symmetry) is, in the two-orbital two-electron description, an "in-phase" zwitterionic state with the electrons paired instantaneously in one or the other nonbonding orbitals of the square. How-

_____

*The idea of the "doubly excited" state of butadience originating in two local triplets coupled into an overall singlet apparently dates back to Linus Pauling before 1940.

ever, just as can the corresponding state in dimethylene [Section 3.4 and (3-26)], this doubly excited state can borrow a significant amount of covalent character. Let us write out its wave function in terms of the lozenged orbitals of the square:

$$S_2\ (^1A_{1g}) \equiv |(a\ -\ c)\ (\overline{a\ -\ c})| \ +\ |(b\ -\ d)\ (\overline{b\ -\ d})| \qquad (5\text{-}18)$$

where a, b, c, and d represent the atomic orbitals at the four corners of the square. Expanding this function, we obtain

$$S_2 \equiv |a\bar{a}| \ +\ |c\bar{c}| \ +\ |b\bar{b}| \ +\ |d\bar{d}| \ -\ (|a\bar{c}| \ +\ |c\bar{a}| \ +\ |b\bar{d}| \ +\ |d\bar{b}|) \qquad (5\text{-}19)$$

in which covalent diagonal terms, from a to c and from b to d, appear. The resonance structure description would be, for all four electrons

$$\hspace{9cm} (5\text{-}20)$$

in which structures carrying adjacent negative charges have been omitted. The diagonally-bonded resonance structure can also be obtained as the sum of two adjacently–bonded covalent structures, as shown by the rectangular orbitals in Table 3-11:*

$$\hspace{9cm} (5\text{-}21)$$

Although the diagonal covalent terms are small and would vanish in the zero-differential–overlap approximation, they do indicate a tendency for *covalent bonding along the diagonals of the square*. Such covalency will be increased by mixture of excited diagonally–bonded covalent configurations of appropriate symmetry, which accounts for the additional stabilization of $S_2\ (^1A_{1g})$ relative to $S_1\ (^1B_{2g})$ (see Table 3-12). In the singlet S_1 the $+$ sign of (5-18) is replaced by a minus sign, which finds its way in front of the two last terms in the parentheses in (5-19). Hence the covalent diagonal terms tend to cancel each other, and the energy of the covalent component is higher. Correspondingly, covalent configurations that might stabilize S_1 have a much higher energy and mix poorly with it.

What, then, are the bonding characteristics of $S_1\ (^1B_{2g}$ symmetry) in the square?

*The author is grateful to P. Karafiloglou for a discussion of the resonance structures of the cyclobutadiene excited singlet.

Michl[15] likens S_1 in H_4 to an excimer, as would be obtained by resonance interaction between the two excited possibilities

$$H_2, H_2^* \leftrightarrow H_2^*, H_2 \tag{5-22}$$

with one resonance mixture being stabilized by Coulomb interaction between the transition dipoles and by charge transfer from one molecule to the other. This interpretation is also inherent to the molecular orbital configuration $(\pi - \pi')$ $(\pi^* + \pi'^*)$ used to describe the S_1 state of two ethylenes (Fig. 5-3). The corresponding wave function

$$S_1 \equiv |(\pi - \pi')(\overline{\pi^* + \pi'^*})| + |(\pi^* + \pi'^*)(\overline{\pi - \pi'})| \tag{5-23}$$

contains both local excitation terms

$$|(\pi\overline{\pi}^* + \pi^*\overline{\pi})| - |(\pi'\overline{\pi'^*} + \pi'^*\,\overline{\pi'})| \tag{5-24}$$

and charge-transfer excitation terms

$$|(\pi\overline{\pi'^*} + \pi'^*\overline{\pi})| - |(\pi'\overline{\pi^*} + \pi^*\overline{\pi'})| \tag{5-25}$$

required to stabilize the excited dimer. Although a minimum that could be attributed to an excimer is not present in the state correlation diagram for ethylene photodimerization,* numerical calculations do indicate[14] a shallow minimum of at least 12 kJ/mole.

5.5 State Correlation Diagrams with Direct Passage from Excited Reactant to Ground Product

Thus far we have considered state correlation diagrams in which there is a 1:1 correlation between reactant and product ground states (albeit through a large barrier in certain reactions) and similar direct connections between reactant excited states and product excited states. In 1972 this author in collaboration[17,18] found a family of reactions with a different and novel behavior: the excited state of reactant correlates directly with ground state of product, and conversely, the ground reactant leads to an excited product state.†

---

*See Ref. 16 for an early prediction of an excimer in the photodimerization of butadiene.

†Although the history of minor discoveries is of little consequence, readers may be interested to know that the history of "surface crossings" dates back to July 1972, at the Baden-Baden IUPAC Photochemistry meeting. This author was particularly intrigued by two lectures, one by Hans Schmid, who discussed the nature of intermediates in the ring-opening of aziridines, and a brilliant one by Gerhard Quinkert, who put forth the detailed sequence of intermediates (using conventional resonance structures) in the photochemistry of cyclohexadienones. What was not clear and required an explanation was which Quinkert resonance structures were on excited surface and which were on ground surface—that is, when

(*continued*)

The method now used differs from that employed to draw Woodward-Hoffmann orbital symmetry diagrams in two respects: by the use of resonance structures rather than orbital configurations and by the choice of a symmetry element—generally the reaction plane—that does not cut through bonds being made or broken as do Woodward-Hoffmann elements. Furthermore, an electron count is used to obtain directly the *state* symmetry. Of course, the method suffers from the fact that, just as a single molecular orbital configuration may not be a good representation of a state (Section 5.1), similarly a single resonance structure need not be an accurate representation, either. Hence symmetries and correlations should be correct, but the detailed description of the states are bound to be relatively crude.

The reaction considered first was the photochemical hydrogen abstraction by ketones or Norrish type II reaction,[21] such as the internal γ-hydrogen abstraction:

$$(5\text{-}26)$$

which is known to arise from a n,π^* singlet or triplet state of the ketone.* After passage through a primary product diradical, two sets of products—a cyclobutanol or an olefin-ketone pair—can be obtained. The crucial step is the first step; the

did deexcitation occur? William Dauben, Nicholas Turro and the author joined in several discussions on the problem and, on the last afternoon of the meeting, decided to collaborate on solving it, with the hope that it would lead ultimately to a classification of photochemical reactions. Turro and Dauben both provided this author with superb overviews of the mechanisms of the main reaction-types respectively in ketone and diene photochemistry. Turro's was on a small slip of paper, written down and given to the author at Baden-Baden, and that delineated the hydrogen abstraction reaction and the α cleavage of ketones in extremely simple terms accessible to a theoretician. By playing over and over again with the resonance structures for the hydrogen abstraction reaction, which seemed the simplest one, the author stumbled across the surface crossing one evening at home in late September 1972. Further information came from Turro in an early October letter in which, with insight, he asked the question concerning α cleavage: "Has this got something to do with crossings of energy surfaces?" Dauben's letter gave a detailed description of diene-triene mechanisms. This letter, which postulated zwitterionic intermediates in the bicyclization of hexatriene, was to be instrumental in leading, two and a half years later, to the sudden polarization effect[19] (whose study would require a specific chapter on excited states).

The first, brief account of the surface crossings was published in France.[17] Prior to that, Zimmerman[20] had used sequences of classical resonance structures, with "continuous electron redistribution," to follow the fate of photochemical reactions. He had noticed that for hydrogen abstraction by ketones and for α cleavage "the excited state proceeds directly to afford the initially observed product species" (a diradical) without "electron demotion," but he failed to draw out the consequences for the potential energy surfaces of reactant and product.

*For n,π^* states, see Ref. 22. The role of n,π^* states in Norrish II photoprocesses was discovered by Kellmann and Dubois.[23]

diradical must be on the ground-state surface of products since it can equally well be obtained by heating the product cyclobutanol [see (4-3)]. Yet the reactant has undergone electronic excitation (sign $h\nu$). What must be elucidated is how this excitation leads to the ground surface. As model we consider the abstraction of a hydrogen atom of methane by the carbonyl group of formaldehyde. The abstraction is assumed to occur in the plane of the carbonyl bond, which remains a plane of symmetry for the entire system throughout the reaction. The resonance structures are then drawn out for ground and excited (n,π^*) states of reactant, and also for ground (diradical) and excited (zwitterionic, see Section 3.1) states of primary product. For each resonance structure, the electron count then involves the carbonyl π electrons, the n electrons on oxygen, and the two electrons that initially make up the CH σ bond:

Reactant ground state $4\sigma,2\pi$

Reactant (n,π^*) excited state $3\sigma,3\pi$

$(5\text{-}27)$

Primary product ground state $3\sigma,3\pi$

Primary product excited state $4\sigma,2\pi$

(in the primary product structures the newly created $p\pi$ lone pair or oxygen must not be forgotten). Clearly, other resonance structures are possible for the excited state of reactant

$(5\text{-}28)$

but this does not change the electron count, nor the symmetry of the state.

The state correlation diagram[17,18] is then drawn out (Fig. 5-8). The electron count serves to specify the total spatial symmetry (A' for $4\sigma,2\pi$ and A'' for the $3\sigma,3\pi$ systems). A *surface crossing* appears, between the two potential-energy sheets of different symmetry, one rising and one descending.

The most noticeable feature is that excited state leads directly downward to product. The extremely facile nature of the photochemical abstraction becomes perfectly understandable. Ab initio calculations[18] qualify only slightly this result:

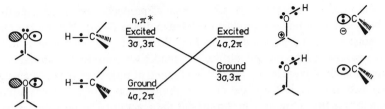

FIGURE 5-8.   State correlation diagram for hydrogen abstraction by ketones.[18] The correlation for excited reactant holds regardless of the spin multiplicity. Reprinted with permission from L. Salem, *J. Am. Chem. Soc.* **96,** 3486 (1974). Copyright 1974 by American Chemical Society.

the calculated surface for n,π^* excited singlet and triplet states in the formaldehyde plus methane reaction are essentially flat rather than descending (Fig. 5-9). Another feature not predicted by the correlation diagram is the small activation energy existing both on the calculated sheets for n,π^* singlet (32 kJ/mole) and for n,π^* triplet (73 kJ/mole) and observed experimentally.[24] The origin of this barrier must be traced back to the "natural" correlation[25] between the n orbital of oxygen and the σ_{OH} orbital, which gives an early dip in the n-orbital energy (Fig. 4-8). Consequently, the n,π^* surface has an early bump.

A second piece of information provided by the state correlation diagram is that reaction in the ground state should lead to an excited-state product and should involve abstraction of a proton. The thermal reaction is not "feasible," as defined by Silver[1]—contrary to both orbital symmetry allowed and orbital symmetry forbidden reactions—since the ground states of reactant and product do not correlate with each other in the presence of the molecular symmetry plane.

Finally, the state correlation diagram allows insight into the distinctive photochemical behavior of states with different multiplicities. The descending nature of the potential sheet for n,π^* excited reactant holds in both singlet and triplet states, and the surface crossing occurs either in $^1A''$ or in $^3A''$ states, whose spatial sym-

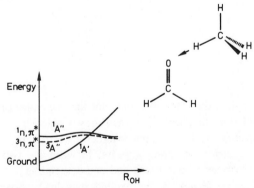

FIGURE 5-9.   Energy surfaces for the formaldehyde + methane hydrogen abstraction reaction.[18] Dotted lines represent triplet energies. Reprinted with permission from L. Salem, *J. Am. Chem. Soc.* **96,** 3486 (1974). Copyright 1974 by American Chemical Society.

metry differs from ground ($^1A'$)-state symmetry. However, the behaviors will *not* be the same if the plane of symmetry is destroyed. The *triplet* surface still correlates directly with product, and its intersection with ground surface remains allowed because of their different *spin* multiplicity. But the singlet excited surface has a well as it curves away from ground surface in an avoided crossing (Section 5.7); a gap then separates the two surfaces. Hence the relative quantum yield of triplet versus singlet is predicted to increase with the nonplanarity of the reaction. Josef Michl* points out that "the gap per se is not detrimental to the quantum yield of the singlet. Rather, it is the fact that the earlier decay—only part-way along the reaction coordinate—to ground singlet S_0 provides an increased opportunity for return to S_0 in the *starting* geometry. In the triplet process the molecule returns to S_0 with the reaction step to biradical completed."

5.6 Nature of State Correlation Diagrams in Relation to Nature of Primary Product

Several other photochemical reactions are found to have state correlation diagrams similar to Fig. 5-8, with a surface crossing between excited singlet state and ground singlet state. This is the case of the photoreduction of azaaromatics, of hydrogen abstraction by singlet carbenes, and of the addition of excited ketones (through their *n* orbital) to electron-rich olefins.[18] Other striking examples include the decomposition of arenediazonium salts—the first calculated surface crossing—and the dissociation of phenol to phenoxy radical.[26]

Other photochemical reactions, however, show entirely different types of correlation diagrams. The α cleavage of ketones (Norrish type I reaction)[21] and cyclic dienones:[27]

$$(5\text{-}29)$$

$$(5\text{-}30)$$

serves to illustrate this point.[18] The electron count technique can again be used to draw out a state correlation diagram, but care is required because the primary product—monoradical in (5-29) or diradical in (5-30)—is not uniquely defined. In both intermediates the odd electron on the carbonyl group may be *either* a σ electron in the σ hybrid left free by rupture of the CC single bond *or* a π electron

*Private communication.

in the relatively low lying π^* orbital of the carbonyl on conjugated systems. In the formyl radical, for instance, the lower-energy situation is the σ radical $^2A'$ state (Fig. 5-10),[28] which has an equilibrium bent structure. The π radical has a linear geometry, where its $^2A''$ state is actually degenerate with the σ radical state ($^2\Pi$, with a "glancing" intersection[29]).

An immediate consequence for the α-cleavage reaction is the existence of two relatively close-lying product states that are energetically in competition. They correspond respectively to σ radical plus departing radical and to π radical plus departing radical. Both situations are diradical situations. There will be a singlet-triplet pair of states for each situation. The state correlation diagram for reaction (5-29), drawn with the same assumptions as in the previous section, is shown in Fig. 5-11. This diagram has novel features relative to that for hydrogen abstraction (Fig. 5-8): (1) the singlet surfaces do not intersect; and (2) the primary product triplet diradical correlates formally with the triplet-excited π,π^* state of the ketone, but originates really from the triplet–excited n,σ^* state, with an oxygen electron having transferred into the σ^* orbital of the breaking bond. In practice, the $^3\pi,\pi^*$ state lies below the $^3n,\sigma^*$ state, so that an avoided crossing—which can be traced back to orbital crossings between n and π, σ^* and π^*—occurs along the descending pathway.[30] The fact that triplet diradical product is alone to correlate with an excited reactant state nicely explains the relatively high reactivity of triplets in α-cleavage reactions. Similarly, there should be two primary product diradicals of different spatial symmetry in (5-30).

We are thus confronted with an entirely new type of correlation diagram. Other correlation diagrams, again with entirely new features, arise if a reaction such as the decomposition of peroxides

$$H_2O_2 \rightarrow 2OH \qquad (5\text{-}31)$$

is considered.[31]

It appears that the main discriminating effects that determine the energetic "patterns" of correlation diagrams are the nature and the number of electronic states in primary product. These are related to the nature and the number of radical sites in primary product. The concept of "topicity" has been introduced to define

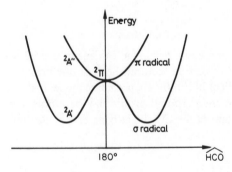

FIGURE 5-10. The two states of the formyl radical as a function of HCO angle bending.[28]

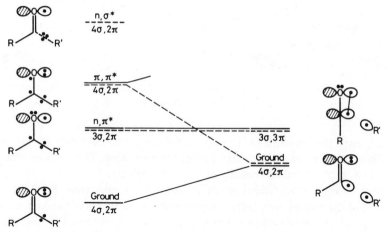

FIGURE 5-11. State correlation diagram for α cleavage of alkanones.[18] Triplet states in dashed lines. For the role of the n,σ^* state, see text.

such a number.[32] The hydrogen abstraction reaction, where two distinct radical sites are created, has a topicity 2. One site (on γ carbon) has σ symmetry, and the other (adjacent to oxygen) has π symmetry and the reaction is said to be "σ,π bitopic." In the α-cleavage reaction the acyl fragment carries two radical sites (σ and π*) and the departing fragment, one site. The reaction is "σ(σ,π) tritopic." The fragmentation of peroxides (5-31) is a reaction with topicity 4, and so forth. The concept can then serve to classify photochemical reactions.[32] Even Woodward-Hoffman photochemically allowed reactions (Section 5.3) are included since the "primary product" antiaromatic cyclobutadienoid transition state is a diradical. The reaction has topicity 2, albeit of a particular type. The most complicated potential surfaces occur for the dissociation of F_2, Cl_2, and so on that possess nine primary product (singlet-triplet) pairs of quasidegenerate states.

5.7 Avoided Crossings Near Real Intersections

We have encountered several examples where the energy surfaces of two states approach each other, as if to intersect and then move apart (Figs. 5-3 and 5-5). In each case a surface intersection that occurred at a simple level of quantum-mechanical calculation disappears at a more sophisticated level, and we speak of an "avoided crossing" (Section 5.1). There are many different types of such avoided crossing.[33]

First, *real* crossings may always occur between surfaces of different spin or spatial symmetry* (we defer the discussion of intersections between surfaces of same symmetry to Section 5-9). If the molecule enters a region of multidimensional space where the symmetry element that permitted the crossing is destroyed, the crossing becomes forbidden, but the "memory" of its existence will stay strong.

---

*For the theory of crossings between electronic states, see Refs. 34–37.

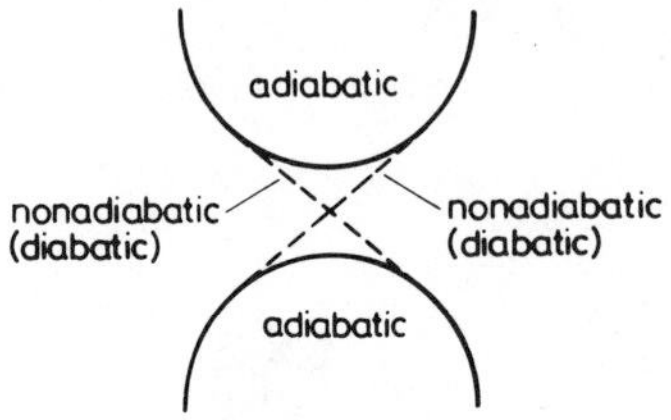

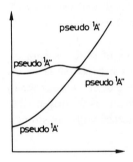

FIGURE 5-12.   Adiabatic (correct) and nonadiabatic (approximate) surfaces.

Indeed, only the presence of a small perturbation (the deviation of the nuclear Hamiltonian from perfect symmetry) deletes the crossing. The surfaces still come very close and an avoided crossing of "type A" occurs.

Other avoided intersections are of an entirely different type (B, C, D, etc.). Because of their simplicity, certain approximate wave functions produce a surface crossing that disappears when the exact wave function and exact Hamiltonian are used. Here there is no physical crossing; the crossing is artificially produced and vanishes when the model is perfected. The "avoided-crossing" terminology is still used because it often coincides with a physically meaningful event.

Regardless of the type of avoided crossing, the states that diagonalize the total electronic Hamiltonian are called *adiabatic;* they are followed by the nuclei if these move slowly enough. The surfaces that give the approximate description and that intersect are called *nonadiabatic* (a pleonastic term) or *diabatic*[38] (Fig. 5-12). The importance of avoided crossings between adiabatic surfaces should not be underscored since they serve in particular as *funnels* through which photoexcited molecules can decay to ground state.[39] This decay is governed by the Landau-Zener law (Section 5.10). Here and in Section 5.8 we consider the different types of avoided crossings,[33] which are best illustrated by drawings.

As mentioned, type A crossings are encountered for molecular geometries close to, but not identical with, a symmetrical geometry for which a crossing occurs rigorously between two electronic states. An example is provided in Fig. 5-13, with the slightly out-of-plane hydrogen abstraction reaction of a ketone. Following adiabatically either the excited or the ground surface there is a symmetry switch at the avoided crossing. The size of the gap between the surfaces can, in a sense, be chosen as a measure of the deviation of the system from C_s symmetry. Another

FIGURE 5-13.   Type A avoided crossing near a true crossing.[33] Compare with Fig. 5-9. The avoided-crossing region is so narrow that the dotted diabatic curves which intersect there can hardly be seen. Reprinted with permission from L. Salem, C. Leforestier, G. Segal, and R. Wetmore, *J. Am. Chem. Soc.* **97**, 479 (1975). Copyright 1975 by American Chemical Society.

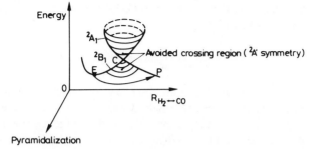

FIGURE 5-14.   Conical intersection in excited H_2CO^+ showing the (type A) avoided crossing region. Ground state is not shown. The pathway from E to P avoids the intersection and uses the avoided-crossing region.

interesting example[40] concerns the photodissociation of H_2CO^+. In the full C_{2v} symmetry the first excited $^2B_1$ and second excited $^2A_1$ states intersect, but this intersection disappears as soon as the molecule pyramidalizes out of planarity. Hence the surface intersection, with a single point in three dimensions, is conical (see Fig. 5-14 and Section 5.9). All around lies the avoided crossing region, in which both surfaces have the same $^2A'$ symmetry, albeit in some regions pseudo-$^2A_1$ and in others pseudo-$^2B_1$.

Now the molecule, after excitation, is on the *lower* surface (at E, say), and the energy barrier required to reach the intersection is a hindrance—contrary to its usefulness in previous photochemical reactions when the molecule was on the *upper* surface of an intersection. But thanks to the out-of-plane bending coordinate, the molecule can use the avoided–crossing region and turn *around* the conical intersection ($E \rightarrow$ product P). In a similar vein thermal abstraction of a CH proton by a ketone could conceivably occur in situations where the nonplanarity would be such as to give a large gap (and a low thermal barrier; see Fig. 5-13) in the avoided-crossing region.

5.8   Avoided Crossings that Are Sequels of Models Too Simple

The other avoided-crossing types originate from the imperfection of various models; the avoided crossing region is often the place of important physical changes along the adiabatic surfaces. Two types arise in the valence-bond model:

Type B

The simplest description of the alkali halides, such as NaCl, uses an ionic wave function similar to (1-6); here

$$\Psi_{ionic} = \phi_{Cl}\,(1)\,\phi_{Cl}\,(2) \qquad (5\text{-}32)$$

where ϕ_{Cl} is the valence $3p$ orbital of chlorine pointing at the sodium atom. The energy curve for this approximate function rises until its dissociation limit

Na^+,Cl^-. A covalent function similar to (1-2) represents quite adequately the lowest excited state of NaCl

$$\Psi_{cov} = \phi_{Cl}(1)\ \phi_{Na}(2) + \phi_{Cl}(2)\ \phi_{Na}(1) \qquad (5\text{-}33)$$

where ϕ_{Na} is the $3s$ orbital of sodium. Contrary to the previous case the energy curve stays relatively flat, going to a radical pair $Na\cdot,Cl\cdot$ limit. Since the energy of this radical pair lies below that of the ion pair, the surfaces of the two functions intersect (Fig. 5-15, dashed lines). When covalent and ionic characters are allowed to mix in each state, as in (1-7), the intersection disappears. At the critical internuclear distance ($R = 10$ Å) where Ψ_{ionic} and Ψ_{cov} are isoenergetic they combine strongly, giving two solutions that are more accurate representations of the real states of the molecule. The lower solution joins smoothly onto the lower state at both shorter and longer bond lengths. The ground surface then leads directly from an ionic molecule to a radical pair (Fig. 5-15), through a short region of mixed character.

The avoided-crossing region is characterized by a spatial "electron jump"[41] from Cl to Na, as the wave function changes abruptly from (5-32) to (5-33). Such an electron jump at an avoided crossing is the major characteristic of electron transfer reactions (Section 8.9). One of the great number of examples has been encountered in Section 3.7 for vinylmethylene.

Type B'

An avoided crossing can occur in valence-bond theory if a system possesses two alternative covalent descriptions, which have opposite energetic behavior along some nuclear coordinate. Hence in the approach of two hydrogen molecules, and for a given geometry, the two resonance structures

$$
\begin{array}{cccc}
\text{H} & \text{H} & \text{H}\!\!-\!\!-\!\!\text{H} & \\
\mapsto & \leftarrow\!\!\mid & \rightarrow & \leftarrow \\
\text{H} & \text{H} & \text{H}\!\!-\!\!-\!\!\text{H} &
\end{array}
\qquad (5\text{-}34)
$$

describe respectively (but approximately) a ground and an excited singlet state. As the molecules approach in the horizontal direction, while increasing the vertical

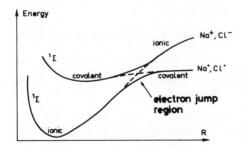

FIGURE 5-15.   Type B avoided crossing.[33] Reprinted with permission from L. Salem, C. Leforestier, G. Segal and R. Wetmore, *J. Am. Chem. Soc.* **97**, 479 (1975). Copyright 1975 by American Chemical Society.

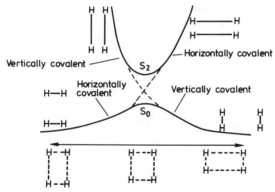

FIGURE 5-16. Type B' avoided crossing: S_2 is described by a sum of the two covalent resonance structures and S_0, by their difference [see also Table 3-11, the rectangular orbital description].

internuclear distance, the energies of the two structures cross*. The crossing becomes avoided,[15] if the wave functions for the two structures are allowed to mix (Fig. 5-16). Note that the higher state S_2 is related to the $^1A_{1g}$ S_2 singlet described for cyclobutadiene in Section 5.4, and that has indeed a contribution from a sum of adjacently–bonded covalent structures as shown in (5-21). Its proper ionic terms (see 5-20) are missing entirely from this valence-bond picture. We now consider avoided crossings that are products of the molecular orbital model.

Type C

This is the case of crossings that become avoided when an independent-electron model is improved by electron repulsion (Section 5.2). The face-on dimerization of ethylene, illustrated in Fig. 5-3, is the most famous example.[6] In the region of the avoided crossing there is a rapid change in molecular orbital configurations [e.g., from $(SS)^2(SA)^2$ to $(SS)^2(AS)^2$ on ground surface]. It should be noted that the root of such an avoided crossing is a *molecular orbital* intersection. This is also true in the type C avoided crossing between excited $^3\sigma\sigma^*$ and $^3\pi\pi^*$ surfaces in the fragmentation of toluene (Fig. 5-17)[42]:

$$C_6H_5CH_3 \rightarrow C_6H_5CH_2^{\cdot} + H^{\cdot} \qquad (5\text{-}35)$$

one of the very first examples studied. A similar avoided crossing is operative between second and third excited triplets in the α cleavage of ketones (Section 5.6).[30]

Type D

Type D avoided crossings are also characterized by an orbital intersection, or near intersection, but between orbitals with different spatial characteristics (rather than

*S. S. Shaik (*J. Am. Chem. Soc.* **103**, 3592 (1981)) has recently generalized the concept that a reaction surface results from a crossing between two valence-bond structures, one describing the old bonds, and the other (excited) the new bonds wih the reactants properly prepared for bonding.

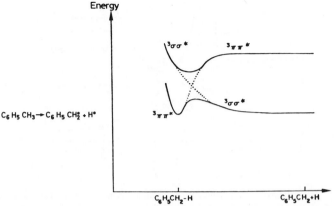

Energy

$C_6H_5CH_3 \rightarrow C_6H_5CH_2^\bullet + H^\bullet$

$^3\sigma\sigma^*$ $^3\pi\pi^*$

$^3\pi\pi^*$ $^3\sigma\sigma^*$

$C_6H_5CH_2$-H $C_6H_5CH_2 + H$

FIGURE 5-17.   Type C avoided crossing between excited states in fragmentation of toluene.[42] Ground state not shown. Reprinted with permission from J. Michl, *Topics Curr. Chem.* **46**, 1 (1974). Copyright 1974 by Springer-Verlag.

different symmetry)—generally a Rydberg orbital and a valence orbital. These orbital intersections arise because Rydberg orbitals are insensitive to geometry changes whereas valence orbitals are not. At the state level, there is an avoided crossing between the flat Rydberg state and the rapidly varying valence state.[43–47] Of course, the avoided-crossing region is characterized by a strong orbital contraction (on one surface, from Rydberg to valence) or orbital expansion (on the other surface, from valence to Rydberg). A first classical example is the coplanar decomposition of excited ammonia[43,47–50]

$$NH_3^* \rightarrow NH_2 + H \cdot \tag{5-36}$$

Here the flat $3s$ Rydberg orbital has a near-intersection with the descending σ_{NH}^* valence orbital; their symmetries are the same, but the Hamiltonian matrix element between them [see (4-6)] is small because one orbital is diffuse and the other relatively contracted. The orbital correlations and state correlations are shown in Fig. 5-18 (the presence of an intermediate state somewhat mars the picture). Other examples of avoided crossings between Rydberg and valence states include the case of BH studied by Mulliken[46] (Fig. 5-18), the photodecomposition of OH_2,[43–45] the avoided crossing between Rydberg $^1\pi,3p_y$ state and valence $^1\pi,\pi^*$ state in the twisting of ethylene (Fig. 3-5),[51] the photodecomposition of CH_4,[52] and many other cases.[47] The transformation of a Rydberg state to a covalent valence state associated with bond twist or bond lengthening has been called "de–Rydbergization."[53]

A not unrelated type of avoided crossing occurs between Rydberg states and ion-pair valence states at very large interatomic distances in the dissociation of diatomics (H_2, Fig. 1-3; Cl_2, etc.) (see Ref. 6 in Chapter 1). The increase in bond length is now accompanied by "Rydbergization"[*] since the ionic valence energy

---

[*]Mulliken[53] speaks of "MO-or-state" Rydbergization, as distinct from cases of "MO" Rydbergization alone, when the molecular orbital changes character but this is not accompanied by any avoided crossing.

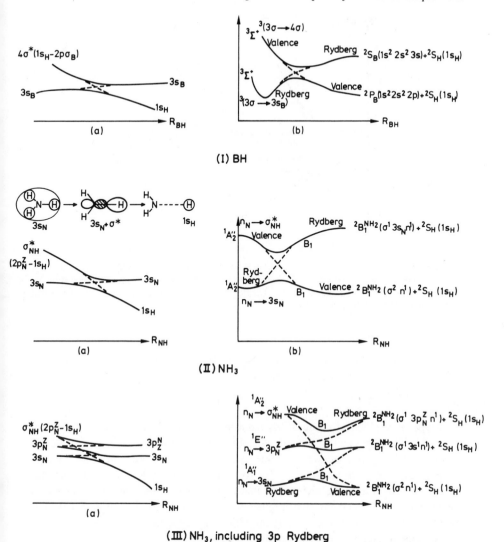

FIGURE 5-18.   Type D avoided crossings: *(a)* orbital correlations; *(b)* state correlations (ground state not shown). For BH, see Ref. 46. For NH_3, see Refs. 50 and 47. In BH orbital 3σ is the bonding combination $1s_H + 2p\sigma_B$; 4σ is their antibonding combination. In NH_3 the simple picture when the sole $3s$ Rydberg orbital is included (II) is marred when the $3p_z$ Rydberg orbital is included (III)[47]. By n_N we mean the $2p_z^N$ orbital perpendicular to the NH_3 plane.

As an example, he gives the $1s\sigma_u$ orbital of H_2, which correlates with $2p\sigma$ of the united atom (compare with $1s\sigma_u$ of $He_2 \rightarrow 2p\sigma$ Be in Fig. 4-1). There is no avoided crossing between orbitals of same symmetry, nor avoided crossing between states of same symmetry. However, Mulliken fails to notice that in this case the change from valence to Rydberg character is accompanied by an orbital crossing between $1s\sigma_u$ and an orbital of different symmetry ($2s\sigma_g \rightarrow 2s$ united atom), and a real state crossing between $^1\Sigma_u^+$ ($1s\sigma_g \, 1s\sigma_u$) and $^1\Sigma_g^+$ ($1s\sigma_g \, 2s\sigma_g$). So a state avoided crossing *or* real crossing always accompanies Rydbergization.

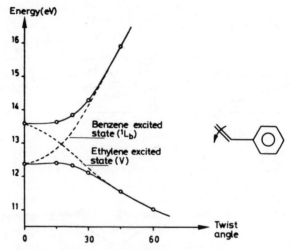

FIGURE 5-19.   Avoided crossing between excited states of styrene localized on the benzene ring (^1L_b) and on the ethylene bond respectively.[55] Ground state not shown. Reprinted with permission from M. C. Bruni, F. Momicchioli, I. Baraldi and J. Langlet, *Chem. Phys. Lett.* **36**, 484 (1975). Copyright 1975 by North-Holland Publishing Company.

increases with distance, contrary to the covalent valence states in the previous cases, which were decreasing in energy.

Type E

This type of avoided crossing arises from approximate descriptions of molecules in terms of local fragments: the energies of the two approximate "fragment states" intersect and this intersection is avoided when fragment interaction is introduced. An example occurs in styrene,[54,55] for rotation around the double bond. The surfaces for excitation on the phenyl and that on the double bond intersect: allowance for interaction between the two fragments destroys the crossing (Fig. 5-19). The avoided crossing is characterized by migration of the excitation from one fragment to the other in either of the two adiabatic states.

5.9   Crossings Between States of Same Symmetry?

The noncrossing rule between states of same symmetry in one dimension (diatomic molecules) was first proved by von Neumann and Wigner.[56,57] It was extended to polyatomic molecules by Teller,[34] with further insight coming from Herzberg and Longuet-Higgins[35,36] and from Carrington.[37] The theory starts with the assumption that two approximate wave functions, Ψ_1 and Ψ_2, have been used to describe the two states. We now allow these two functions to combine and interact through the full Hamiltonian H, or whatever part of the Hamiltonian has been omitted initially.

The combination

$$\Psi = c_1\Psi_1 + c_2\Psi_2 \tag{5-37}$$

leads, after applying the variational principle, to the secular determinantal equation (see also 5-6)

$$\begin{vmatrix} H_{11} - E & H_{12} \\ H_{21} & H_{22} - E \end{vmatrix} = 0 \tag{5-38}$$

Since $H_{12} = H_{21}$, the roots of the quadratic equation

$$E = \frac{1}{2}\{H_{11} + H_{22} \pm \sqrt{(H_{11} - H_{22})^2 + 4H_{12}^2}\} \tag{5-39}$$

are equal only if the quantity under the square root vanishes. Equal energies, and hence the existence of an intersection, therefore requires *two* conditions:

$$H_{11} = H_{22} \tag{5-40}$$
$$H_{12} = 0$$

We now analyze these conditions successively for one nuclear coordinate (the energy plot is then a two-dimensional one), two coordinates (energy plot in three dimensions), or n coordinates (energy plot in $n + 1$ dimensions).

With only one variable, the internuclear distance, it is generally impossible to satisfy simultaneously two conditions. So, unless $H_{12} = 0$ *by symmetry,* the energy roots are different. In one dimension two states of same symmetry cannot cross.[56,57] This should also be true for polyatomic molecules if a reaction coordinate in which all variables are constrained to vary linearly together is used—since the problem is really reduced to a single variable.

If there are two nuclear coordinates Q_1 and Q_2, there should be a value Q_1^0, Q_2^0 of these for which the conditions (5-40) are fulfilled. Choosing this point as origin, the diagonal $H_{11} - H_{22}$ term and the off-diagonal matrix element H_{12} will be linear in small increments of the nuclear coordinates near it (Taylor expansion):

$$H_{12} = aQ_1 + bQ_2$$

$$\frac{b}{a} = \frac{d}{c} = \frac{Q_1^0}{Q_2^0} \tag{5-41}$$

$$H_{11} - H_{22} = cQ_1 + dQ_2$$

It is always possible to transform the coordinates Q_1 and Q_2 to x and y such that H_{12} is a function of y alone and such that $H_{11} + H_{22}$ vanishes[34]. Equation (5-39) then becomes the equation for a *double cone.* The simplest such example of conical intersections arises in linear triatomic molecules, such as HNO, in which the lowest $^1\Delta$ state (configuration $\sigma^2\pi^4\pi^{*2}$; compare with O_2 in Section 3.3) intersects a higher $^1\Pi$ state (configuration $\sigma^2\pi^4\pi^*\sigma^*$) (Fig. 5-20). The higher $^1\Pi$ state, however, dissociates into a hydrogen atom and ground state NO ($^2\Pi$, $\sigma^2\pi^4\pi^*$), whereas the $^1\Delta$ state dissociates to a hydrogen atom plus a highly excited NO ($^2\Delta$, $\sigma\pi^4\pi^{*2}$ $\leftrightarrow$ $\sigma^2\pi^3\pi^*\sigma^*$).

If the molecule is bent out of a straight line along the y coordinate at crossing point C, then (1) each state splits into a symmetric A' component and an antisymmetric A'' component with respect to the plane of bending and (2) the nuclear perturbation, which is symmetrical with respect to this plane, mixes the two A' components; $H_{12} \neq 0$, and the crossing becomes avoided. At point C:

$$
\begin{array}{lcl}
^1\Delta \quad \xrightarrow{\hspace{2cm}} & \begin{cases} ^1A'' \\ ^1A' \end{cases} & \text{-----------} \rightarrow \quad ^1A' \\[1em]
^1\Pi \quad \xrightarrow{\hspace{2cm}} & \begin{cases} ^1A' \\ ^1A'' \end{cases} & \text{-----------} \rightarrow \quad ^1A'
\end{array}
$$

Equal energies   Symmetry reduced to C_s Bending   Different energies   (5-42)

A characteristic property of such a conical intersection is that the wave function changes sign[35] when

$$x,y \rightarrow -x, -y \tag{5-43}$$

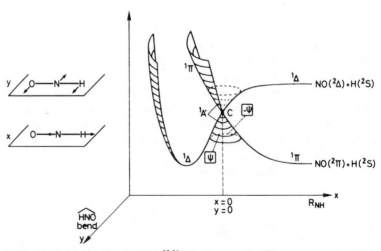

FIGURE 5-20.   Conical intersection in HNO.[35,36] The wave function Ψ changes sign on opposite signs of the cone. A similar intersection occurs for the A'' surfaces (not shown). Adapted from *Electronic Spectra of Polyatomic Molecules* by G. Herzberg. Copyright 1966 by Van Nostrand Reinhold Company. Reprinted by permission of the publisher.

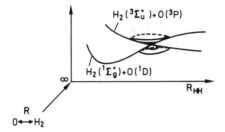

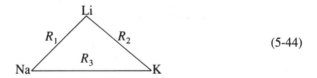

FIGURE 5-21. Half-conical intersection in H_2O.[3,36] Ground-state surface H_2 ($^1\Sigma_g^+$) + O(^{3}P) not shown.

(Fig. 5-20). Another such intersection occurs in OH_2, in the unlikely geometry where the oxygen atom is infinitely removed from the hydrogens (Fig. 5-21).[3,36] It is really a half-conical intersection since the y coordinate cannot go beyond infinity. We have also encountered the case of H_2CO^+ (Section 5.7).

In all these cases, however, the intersection point arises from *symmetry;* thus these are really extensions to two internuclear coordinates of the allowed intersection betweeen states of different symmetry for a single coordinate (diatomics). Yet in theory (5-40) can be satisfied without the help of symmetry even for two nuclear coordinates, and for n coordinates, one should be able to satisfy (5-40) in $n - 2$ dimensions. Therefore, Teller[34] and Herzberg and Longuet-Higgins[35] concluded that two states *even with same symmetry* will intersect along a $(n - 2)$-dimensional hyperline as the energy is plotted against the n nuclear coordinates. In the simplest case of three nuclear coordinates, the energy plot should yield an intersection point. An example is the triangular molecule

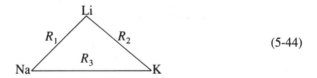

$$(5\text{-}44)$$

in which the two would-be intersecting states have $^2A'$ symmetry (both correspond to the degenerate $^2E'$ pair of states in triangular H_3). Naqvi[58] and Hoytink[59] strongly contested this point and suggested that a different symmetry alone can enforce the crossing. Taking a London-Eyring-Polanyi energy formula (1-35) for the double cone:

$$E = Q_{AB} + Q_{BC} + Q_{CA}$$

$$\pm \left[\frac{1}{2} (J_{AB} - J_{BC})^2 + \frac{1}{2} (J_{BC} - J_{CA})^2 + \frac{1}{2} (J_{CA} - J_{AB})^2 \right]^{1/2} \quad (5\text{-}45)$$

the question becomes whether the square root can vanish in a heteronuclear triangular triatomic or whether it is zero only when all atoms are identical:

$$J_{AB} = J_{BC} = J_{CA} \quad (5\text{-}46)$$

in which case different symmetries would again be present. A strong indication that (5-46) can be satisfied independently of symmetry can be obtained by showing both analytically[60] and numerically[61] that the wave functions change sign when

$$\frac{1}{\sqrt{2}} (R_2 - R_2^e) + \frac{1}{\sqrt{2}} (R_3 - R_3^e) = x \to -x$$

(5-47)

$$\sqrt{\frac{2}{3}} (R_1 - R_1^e) - \frac{1}{\sqrt{6}} (R_2 - R_2^e) - \frac{1}{\sqrt{6}} (R_3 - R_3^e) = y \to -y$$

a distinctive feature of conical intersections. We propose a different qualitative proof of (5-46) as follows (Fig. 5-22). Assume A to be small, C large, and B medium size. Draw out bond AC at its assumed equilibrium distance (the corresponding value of the exchange integral is J_{AC}). We seek the position of B. Around A draw a circle with an AB bond length such that

$$J_{AB} = J_{AC}$$

(5-48)

Since B is smaller than C, this bond length should be smaller than R_{AC}. Similarly, around C draw a circle with a CB bond length corresponding to

$$J_{BC} = J_{AC}$$

(5-49)

Since B is larger than A, the bond length is larger than R_{AC}. The two circles necessarily intersect, yielding the locus of point B that satisfies (5-46). The corresponding geometry at the surface intersection has

$$R_{BC} > R_{AC} > R_{AB}$$

(5-50)

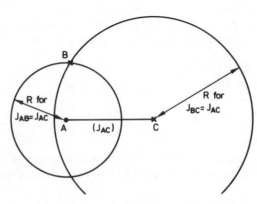

FIGURE 5-22.   Finding the geometry of the conical intersection in a heteronuclear triatomic ABC.

which agrees nicely with that found for potassium (large), sodium (medium), and lithium (small) atoms.*

Despite these demonstrations, it appears that a good part of the $(n - 2)$-dimensional hyperline of intersection between surfaces of same symmetry may be at infinity or even in the imaginary plane. Recently the conditions for crossing between two states in a three-state system have been described.† All descriptions insist on the rarity of these crossings.

5.10 Landau-Zener Law

What usefulness does the concept of an avoided crossing have? If the correct curves are ultimately the adiabatic ones, which avoid each other, what profit can be made from knowing that in some approximation the surfaces would cross? The answer is that there is an important physical connection between the two diagonally opposed branches of two avoiding surfaces (upper left, lower right, say, which connect diabatically; see Fig. 5-12): *they have nearly the same electronic wave function.* This wave function, on the other hand, changes violently along the adiabatic surface in the avoided-crossing region. This can be verified in all the examples given in Figs. 5-12–5-18, including the type A case, where the electronic spatial symmetry characteristics are preserved along the diabatic surfaces and destroyed along the adiabatic ones (Fig. 5-13). Insofar as electrons determine the potential energy to which nuclei are ascribed, the nuclei tend to follow the diabatic surfaces for which the electronic wave function and "electron field" remains the same—unless they go sufficiently slowly to feel the changing potential field imposed by the adiabatic surfaces and to be "captured" by it.

What, then, is the probability that the nuclei will remain on a diabatic surface and cross, for instance, from excited to ground state (e.g., as in Figs. 5-3, 5-6, and 5-12)? The answer was given by Landau and Zener.[66] This probability, which is also that for decay from one adiabatic surface to the other, is, as a first approximation, given by

$$P = \exp\frac{-\pi^2 g^2}{\hbar v \delta s} \tag{5-51}$$

where v is the nuclear velocity along the reaction coordinate, δs is the difference in slopes between the intersecting diabatic states, and g is the energy gap at the avoided crossing. An illustration of the role of these three factors has been given by Devaquet.[33]

The factor g, the energy gap between the two *adiabatic* states at the crossing point $(H_{11} = H_{22})$, is, from (5-39)

---

*The geometry of KNaLi for the assumed intersection is obtained in Fig. 1 of Varandas et al.[61] for $\theta = 0$.

†See Refs. 62 and 63, which partly answer a paper by Hatton et al.,[64] who discuss spurious avoided crossings. See also Ref. 65.

$$g = 2H_{12} \qquad (5\text{-}52)$$

a relation that we have already studied in type C avoided crossings when $H_{12} = K_{12}$ (Section 5.1). A large gap implies a strong interaction between the diabatic states; hence the wave functions leading to the crossing are poor and the true wave functions deviate strongly from them, even far away from the crossing. It is hard for the nuclei to follow the path dictated by the approximate functions and they tend to remain on the adiabatic surfaces. Hence the larger g, the smaller P.

Consider now two situations characterized by two different values of δs (Fig. 5-23). In one case, with a small difference in slopes, the variation in wave function to which the nuclei must adapt—if they are to remain on an adiabatic surface—operates over a large region of coordinate space. The nuclei have time to ''see'' the changes in electronic energy and remain on an adiabatic surface; P is small again. When δs is large, the changes in electronic energy are only ''seen'' by the nuclei through a small window in coordinate space. The probability of decay is large. The role of the velocity v is parallel to that of δs: rapid passage through the crossing region ensures that the nuclei can ignore variations in the electronic wave function and its potential; the larger v, the larger P.

The Landau-Zener formula, however, is a semiclassical one. It gives a decay probability, but no evaluation of a decay rate. Such a decay rate was calculated by Labhart[67] a physical chemist of great modesty and remarkable talent, who died 2 weeks after submitting this last work. From a purely quantum-mechanical viewpoint, passage from one adiabatic surface with exact electronic wave function ψ_+ to another adiabatic surface with exact electronic wave function ψ_-, where ψ_+ and ψ_- are the two roots of (5-38) in terms of the diabatic functions ψ_1 and ψ_2^{*}

$$\psi_+ = \cos\alpha \, (\psi_1) + \sin\alpha \, (\psi_2)$$
$$\psi_- = \sin\alpha \, (\psi_1) - \cos\alpha \, (\psi_2) \qquad (5\text{-}53)$$

(a) **δs** small   (b) **δs** large

FIGURE 5-23.   Small *(a)* and large (b) difference in the slopes of the diabatic (intersecting) potential energy surfaces in the case of an avoided crossing (Devaquet[33]). Reprinted with permission from A. Devaquet, *Pure Appl. Chem.* **41**, 455 (1975). Copyright 1975 by International Union of Pure and Applied Chemistry.

---

*Exceptionally we write ψ for electronic states and keep capital Ψ for vibronic states. The sign χ describes the nuclear vibrational wave functions.

can come only from a breakdown in the Born-Oppenheimer approximation (Section 1.1). The "non–Born-Oppenheimer" term is the matrix element of nuclear coordinate differentials arising from the nuclear kinetic energy operator:

$$b = \langle \psi_+ \chi_+ | \mathcal{H}_{\text{non–Born-Oppenheimer}} | \psi_- \chi_- \rangle \tag{5-54}$$

$$= \langle \psi_+ \chi_+ | \sum_j - \frac{\hbar^2}{2\mu_j} \left(\frac{\partial^2 \psi_-}{\delta Q_j^2} \chi_- + 2 \frac{\partial \psi_-}{\delta Q_j} \frac{\partial \chi_-}{\delta Q_j} \right) \rangle$$

between "vibrational-plus-electronic" or "vibronic" wave functions. In (5-54) μ_j is the mass of the jth nucleus of coordinate Q_j. The existence of such a matrix element ensures that there is a connection between upper and lower adiabatic states. The correct stationary vibronic states will be linear combinations.

$$\Psi' = \cos\beta(\psi_+ \chi_+) + \sin\beta(\psi_- \chi_-)$$
$$\Psi'' = -\sin\beta(\psi_+ \chi_+) + \cos\beta(\psi_- \chi_-) \tag{5-55}$$

where the new energies E' and E'' are related to the adiabatic energies E_+ and E_- by

$$E' - E'' = \frac{E_+ - E_-}{\cos(2\beta)} = \frac{2b}{\sin(2\beta)} \tag{5-56}$$

The general time-dependent *nonstationary* state of the system can now be represented in terms of the new diabatic basis*

$$\Psi = A'(t)e^{-i(E' - E'')t/(2\hbar)} \Psi' + A''(t)e^{i(E' - E'')t/(2\hbar)} \Psi'' \tag{5-57}$$

or better even, through (5-55) and (5-56), in terms of the original adiabatic basis:

$$\Psi = \{A'(t) \cdot e^{-ibt/[\hbar \sin(2\beta)]} \cdot \cos\beta - A''(t) \cdot e^{ibt/[\hbar \sin(2\beta)]} \cdot \sin\beta\} \, \psi_+ \chi_+$$
$$+ \{A'(t) \cdot e^{-ibt/[\hbar \sin(2\beta)]} \cdot \sin\beta + A''(t) \cdot e^{ibt/[\hbar \sin(2\beta)]} \cdot \cos\beta\} \, \psi_- \chi_- \tag{5-58}$$

Equation (5-58) shows that, if the system at time $t = 0$ is on ψ_+, at time $t = \hbar\pi \sin(2\beta)/(2b)$ it will be on ψ_-. Thus the rate constant for passage from one adiabatic state to the other is of the order

$$\mathbf{k} \approx \frac{b}{\hbar} \tag{5-59}$$

For model cases ($\delta s \approx 2$ eV/Å, $\mu = 1.67 \times 10^{-24}$ g, $H_{12} = 0.02$ eV) Labhart finds

*The author is grateful to X. Chapuisat for a discussion of this problem.

$$\mathbf{k} \approx 10^{13} \text{ s}^{-1} \qquad (5\text{-}60)$$

Above a critical gap width ($2H_{12} = 0.2$ eV) the matrix element b decreases steeply over many orders of magnitude, so that for a gap of 1 eV the decay rate has fallen to 1 (s)$^{-1}$!

The presence of a second, higher surface may modify the normal Arrhenius behavior (Section 2.2) on a ground surface at an avoided crossing.[68] The effective energy barrier is found to depend on H_{12}, and the preexponential factor has an additional temperature-dependent component.

References

1. D. M. Silver, *J. Am. Chem. Soc.* **96,** 5959 (1974).
2. E. Wigner and E. E. Witmer, *Z. Physik* **51,** 859 (1928).
3. K. E. Shuler, *J. Chem. Phys.* **21,** 624 (1953).
4. T. F. George and J. Ross, *J. Chem. Phys.* **55,** 3851 (1971).
5. O. Sinanoglu, *J. Chem. Phys.* **36,** 706, 3198 (1962).
6. H. C. Longuet-Higgins and E. W. Abrahamson, *J. Am. Chem. Soc.* **87,** 2045 (1965).
7. J. S. Wright and L. Salem, *J. Am. Chem. Soc.* **94,** 322 (1972).
8. R. W. Carr and W. D. Walters, *J. Phys. Chem.* **67,** 1370 (1963); H. M. Frey, *Adv. Phys. Org. Chem.* **4,** 147 (1966).
9. W. T. A. M. van der Lugt and L. J. Oosterhoff, *J. Am. Chem. Soc.* **91,** 6042 (1969).
10. J. Koutecky, *J. Chem. Phys.* **47,** 1501 (1967); K. Schulten and M. Karplus, *Chem. Phys. Lett.* **14,** 305 (1972); S. Shih, R. J. Buenker, and S. Peyerimhoff, ibid. **16,** 244 (1972); T. H. Dunning, R. P. Hosteny, and I. Shavitt, *J. Am. Chem. Soc.* **95,** 5067 (1973).
11. R. B. Woodward and R. Hoffmann, *The Conservation of Orbital Symmetry* (Verlag Chemie, Weinheim, 1970), p. 176, footnote 253.
12. D. Grimbert, G. Segal, and A. Devaquet, *J. Am. Chem. Soc.* **97,** 6629 (1975).
13. W. G. Dauben, R. L. Cargill, R. M. Coates, and J. Saltiel, *J. Am. Chem. Soc.* **88,** 2742 (1966).
14. E. Kassab, E. M. Evleth, J. J. Dannenberg, and J. C. Rayez, *Chem. Phys.* **52,** 151 (1980).
15. J. Michl, *Pure Appl. Chem.* **41,** 507 (1975); W. Gerhartz, R. D. Poshusta, and J. Michl, *J. Am. Chem. Soc.* **98,** 6427 (1976).
16. L. Salem, *J. Am. Chem. Soc.* **90,** 543, 553 (1968).
17. L. Salem, W. G. Dauben, and N. J. Turro, *J. Chim. Phys. Physiochim. Biol.* **70,** 694 (1973).
18. L. Salem, *J. Am. Chem. Soc.* **96,** 3486 (1974).
19. V. Bonacic-Koutecky, P. Bruckmann, P. Hiberty, J. Koutecky, C. Leforestier, and L. Salem, *Angew. Chem. Internatl. Ed.* **14,** 575 (1975); L. Salem, *Acc. Chem. Res.* **12,** 87 (1979).
20. H. E. Zimmerman, *Abstracts, 17th National Organic Symposium,* Bloomington, Indiana, June 1961, p. 31; *Adv. Photochem.* **1,** 183 (1963); *Science* **153,** 837 (1966); H. E. Zimmerman and D. I. Schuster, *J. Am. Chem. Soc.* **83,** 4486 (1961); ibid. **84,** 4527 (1962).
21. R. G. W. Norrish, *Transact. Faraday Soc.* **33,** 1521 (1937).
22. M. Kasha, *Radiat. Res.,* **Suppl. 2,** 243 (1960); M. Kasha in *Light and Life* edited by W. D. McElroy and B. Glass (John Hopkins University Press, Baltimore, 1961), p. 31.
23. A. Kellmann and J. T. Dubois, *J. Chem. Phys.* **42,** 2518 (1965).
24. F. D. Lewis, *Molec. Photochem.* **4,** 501 (1972); J. C. Scaiano, J. Grotewold, and C. M. Previtali, *J. Chem. Soc. Chem. Commun.* 350 (1972).
25. A. Devaquet, A. Sevin, and B. Bigot, *J. Am. Chem. Soc.* **100,** 2009 (1978).
26. R. J. Cox, P. Bushnell, and E. M. Evleth, *Tetrahedron Lett.* 207 (1970); E. M. Evleth, P. M. Morowitz, and T. S. Lee, *J. Am. Chem. Soc.* **95,** 7948 (1973).
27. G. Quinkert, *Pure Appl. Chem.* **33,** 285 (1973); O. L. Chapman and J. D. Lassila, *J. Am. Chem. Soc.* **90,** 2449 (1968); G. Quinkert, F. Cech, E. Kleiner, and D. Rehm, *Angew. Chem. Internatl. Ed.* **18,** 557 (1979).

28. J. W. C. Johns, S. H. Priddle, and D. A. Ramsay, *Disc. Faraday Soc.* **35**, 90 (1963); G. Herzberg, *Electronic Spectra and Electronic Structure of Polyatomic Molecules* (Van Nostrand, Princeton, 1966), p. 469; K. Tanaka and E. R. Davidson, *J. Chem. Phys.* **70**, 2904 (1979).
29. T. Carrington, *Disc. Faraday Soc.* **53**, 27 (1972).
30. A. Sevin, B. Bigot, and A. Devaquet, *Tetrahedron* **34**, 3275 (1978).
31. E. M. Evleth, *J. Am. Chem. Soc.* **98**, 1632 (1976); E. M. Evleth and E. Kassab, ibid. **100**, 7859 (1978).
32. W. G. Dauben, L. Salem, and N. J. Turro, *Acc. Chem. Res.* **8**, 41 (1975).
33. L. Salem, C. Leforestier, G. Segal, and R. Wetmore, *J. Am. Chem. Soc.* **97**, 479 (1975); A. Devaquet, *Pure Appl. Chem.* **41**, 455 (1975).
34. E. Teller, *J. Phys. Chem.* **41**, 109 (1937).
35. G. Herzberg and H. C. Longuet-Higgins, *Transact. Faraday Soc.* **35**, 77 (1963).
36. G. Herzberg, *The Electronic Spectra of Polyatomic Molecules* (Van Nostrand, Princeton, 1966), p. 442.
37. T. Carrington, *Disc. Faraday Soc.* **53**, 27 (1972).
38. T. F. O'Malley, *Adv. Atom. Molec. Phys.* **7**, 223 (1971).
39. J. Michl, *Molec. Photochem.* **4**, 243 (1972).
40. J. C. Lorquet, A. J. Lorquet, and M. Desouter-Lecomte in *Quantum Theory of Chemical Reactions*, Vol. II, edited by R. Daudel, A. Pullman, L. Salem, and A. Veillard (Reidel, Dordrecht, 1981), p. 241.
41. J. L. Magee, *J. Chem. Phys.* **8**, 687 (1940); R. Grice and D. Herschbach, *Molec. Phys.* **27**, 159 (1974).
42. J. Michl, *Topics Curr. Chem.* **46**, 1 (1974).
43. J. A. Horsley and F. Flouquet, *Chem. Phys. Lett.* **5**, 165 (1970).
44. J. A. Horsley and W. H. Fink, *J. Chem. Phys.* **50**, 750 (1969).
45. F. Flouquet and J. A. Horsley, *J. Chem. Phys.* **60**, 3767 (1974).
46. R. S. Mulliken, *Internatl. J. Quant. Chem.* **55**, 83 (1971).
47. E. M. Evleth and E. Kassab in *Quantum Theory of Chemical Reactions,* Vol. II, edited by R. Daudel, A. Pullman, L. Salem, and A. Veillard (Reidel, Dordrecht, 1981), p. 261; E. Kassab, J. T. Gleghorn, and E. M. Evleth, *Chem. Phys. Lett.* **70**, 151 (1980).
48. A. E. Douglas, *Disc. Faraday Soc.* **35**, 158 (1963).
49. G. Herzberg, *The Electronic Spectra of Polyatomic Molecules* (Van Nostrand, Princeton, N.J., 1966), p. 465.
50. R. Runau, S. D. Peyerimhoff, and R. J. Buenker, *J. Molec. Spectr.* **68**, 253 (1977).
51. S. D. Peyerimhoff and R. J. Buenker, *Theoret. Chim. Acta (Berl.)* **27**, 243 (1972); R. J. Buenker and S. D. Peyerimhoff, *Chem. Phys.* **9**, 75 (1976).
52. M. S. Gordon, *Chem. Phys. Lett.* **52**, 161 (1977); M. S. Gordon and J. W. Caldwell, *J. Chem. Phys.* **70**, 5503 (1979).
53. R. S. Mulliken, *Acc. Chem. Res.* **9**, 7 (1976); *Chem. Phys. Lett.* **46**, 197 (1977).
54. M. G. Rockley and K. Salisbury, *J. Chem. Soc., Perkin Transact.* 1582 (1973).
55. M. C. Bruni, F. Momicchioli, I. Baraldi, and J. Langlet, *Chem. Phys. Lett.* **36**, 484 (1975).
56. J. von Neumann and E. Wigner, *Physik. Z.* **30**, 467 (1929).
57. C. A. Coulson, *Valence,* 2nd ed. (Oxford University Press, Oxford, 1961), p. 65.
58. K. R. Naqvi, *Chem. Phys. Lett.* **15**, 634 (1972).
59. G. J. Hoytink, *Chem. Phys. Lett.* **34**, 414 (1975).
60. H. C. Longuet-Higgins, *Proc. Roy. Soc.* **A344**, 147 (1975).
61. A. J. C. Varandas, J. Tennyson, and J. N. Murrell, *Chem. Phys. Lett.* **61**, 431 (1979).
62. C. M. Meerman-van Benthem, A. H. Huizer, and J. J. C. Mulder, *Chem. Phys. Lett.* **51**, 93 (1977).
63. N. Moiseyev, G. F. Kventsel, and J. Katriel, *Chem. Phys. Lett.* **57**, 477 (1978).
64. G. J. Hatton, W. L. Lichten, and N. Ostrove, *Chem. Phys. Lett.* **40**, 437 (1976).
65. T. Carrington, *J. Chem. Phys.* **70**, 2958 (1979).
66. L. Landau, *Phys. Z. Sowjet.* **2**, 46 (1932); C. Zener, *Proc. Roy. Soc.* **A137**, 696 (1932); E. C. G. Stueckelberg, *Helv. Phys. Acta* **5**, 369 (1933).
67. H. Labhart, *Chem. Phys.* **23**, 1 (1977).
68. J. P. Laplante and W. Siebrand, *Chem. Phys. Lett.* **59**, 433 (1978).

6

Reaction Paths
from the Properties of Reactants

The notion that molecules contain all the information necessary to understand their reactions with other molecules has long been dear to chemists. The term "reactivity" carries the implications that the ability of a molecule to react is predetermined by its own structural properties. Common sense and experience tell us that this cannot be more than a half-truth. Reactivity must depend on the reaction partner, and the encounter of two molecules adds unique features and behavior that neither molecule possesses alone. In the same vein, the manner in which molecules A and B react together, and also B and C together, are not generally sufficient to tell us how A and C will react together. Despite these shortcomings, the idea that reaction paths (unimolecular reactions, or bimolecular reactions with partners of a given family) can be determined by features of single molecules has been a fruitful concept. Better still, an "adequate fit" of certain molecular properties belonging to two reaction partners, by partially accounting for the unique feature mentioned above, has proved to be extremely useful.

6.1 Introduction: "Perturbation" Approach to Reaction Paths

For the reactant, the incipient interactions with a foreign molecule or the initial nuclear displacements in a unimolecular deformation are initially small perturbations. To such small perturbations should correspond small energy changes, which can be evaluated by perturbation expansions of the energy. This perturbation energy is expected to give a reliable indication of the energy in the initial stages of the reaction. We write for the total Hamiltonian:

$$H = H_0 + H' \qquad (6-1)$$

where the "perturbation" Hamiltonian is

$$H' = \begin{cases} H^{\text{int}} & \text{(bimolecular)} & (6\text{-}2a) \\ \\ \Delta(V_{nn} + V_{ne}) & \text{(unimolecular)} & (6\text{-}2b) \end{cases}$$

In (6-2a) the term H^{int} represents the interaction Hamiltonian between the two approaching molecules; the unperturbed Hamiltonian H_0 is then the sum of the

158

Hamiltonians for the separate molecules. In (6-2b) V_{nn} and V_{ne} are respectively the nuclear-nuclear and nuclear-electronic Coulombic potential operators in the distorting molecule.

In the first case the calculation is complicated by the fact that the unperturbed "zero-order" wave functions are simple products of the wave functions of the two molecules

$$
\begin{aligned}
\Psi^0 &= \Psi_0^A \, \Psi_0^B \quad \text{(energy } E_0) \\
\Psi^* &= \Psi_i^A \, \Psi_j^B \quad \text{(energy } E_*)
\end{aligned}
\tag{6-3}
$$

(0, ground, $*$, excited). In the presence of the perturbation, however, the correct wave functions must be antisymmetrized with respect to the exchange of electrons *between* the molecules

$$
\begin{aligned}
|\Psi^0| &= |\Psi_0^A \Psi_0^B| \quad \text{(energy } \overline{E}_0) \\
|\Psi^*| &= |\Psi_i^A \Psi_j^B| \quad \text{(energy } \overline{E}_*)
\end{aligned}
\tag{6-4}
$$

The correct theory of such interactions has been given by Claverie.[1,2,*] In practice, the perturbational expansion of the wave function is done in terms of the fully antisymmetrized functions of type (6-4). The perturbation energy is then given by the sum of a "first-order" term E' and of a "second-order" term E''

$$
E' = H_{00}^{\text{int}}/S_{00}
\tag{6-5a}
$$

$$
E'' = -\frac{1}{S_{00}} \sum_{\substack{\text{excited} \\ \text{states}*}} \frac{|H_{0*}^{\text{int}} - S_{0*}\overline{E}_0|^2}{(\overline{E}_* - \overline{E}_0)S_{**}}
\tag{6-5b}
$$

where the subscripts 0 and $*$ now denote respectively antisymmetrized functions for the combined system of energy $\overline{E}_0$ and $\overline{E}_*$. Excitations in (6-5b) include both local excitations on the reactants (singly or together) and charge-transfer excitations from one side to the other or simultaneously two-way. The interpretation of the first-order term E' is given in Section 6.2; here we concentrate on E''.

We assume the Slater determinants (6-4) to be normalized ($S_{00} = S_{**} = 1$); this is not strictly true because Ψ_i^A and Ψ_j^B are not orthogonal at finite distances. The error is of the order of the square of the intermolecular overlap and varies in size with the nature of the excited state, being larger and larger for higher excited states. Therefore, to be valid, this approximation requires that the basis set of states be finite.[2]

At large distances (6-5b) for the second-order term E'' gives rise to weak intermolecular forces such as dispersion forces, polarization forces, and so on. In the chemical range, the major contribution to (6-5b) generally comes from singly ex-

---

*One of the pioneering papers in this field is that by Murrell et al.[3]

cited states Ψ^* corresponding to charge transfer from one fragment to the other. In molecular orbital language, to the ground state

$$\Psi_{00} = |\psi_1^A \overline{\psi_1^A} \cdots \psi_i^A \overline{\psi_i^A} \cdots \psi_m^A \overline{\psi_m^A} \psi_1^B \overline{\psi_1^B} \cdots \psi_m^B \overline{\psi_m^B}| \qquad (6\text{-}6)$$

corresponds the charge-transfer state:

$$
\begin{aligned}
\Psi_{0,i_A \rightarrow p_B}^* = \frac{1}{\sqrt{2}} \Big[&|\psi_1^A \overline{\psi_1^A} \cdots \psi_i^A \cdots \psi_m^A \overline{\psi_m^A} \; \psi_1^B \overline{\psi_1^B} \cdots \psi_m^B \overline{\psi_m^B} \cdots \overline{\psi_p^B}| \\
- &|\psi_1^A \overline{\psi_1^A} \cdots \overline{\psi_i^A} \cdots \psi_m^A \; \overline{\psi_m^A} \; \psi_1^B \overline{\psi_1^B} \cdots \psi_m^B \overline{\psi_m^B} \cdots \psi_p^B| \Big]
\end{aligned}
\qquad (6\text{-}7)
$$

The necessity of such charge transfer seems intuitively correct since bonding between two reactants must require some form of electron flow between them. Restricting the summation to states of type (6-7), we obtain

$$E'' = - \sum_{\substack{\text{Occupied} \\ \text{orbitals} \\ i \text{ of A}}} \sum_{\substack{\text{Unoccupied} \\ \text{orbitals } p \\ \text{of B}}} \frac{|\sqrt{2} \langle \psi_i^A | h^{\text{int}} | \psi_p^B \rangle - \sqrt{2} \langle \psi_i^A | \psi_p^B \rangle E_0|^2}{\overline{E}_{i_A \rightarrow p_B} - \overline{E}_0} \qquad (6\text{-}8)$$

+ similar term with roles of A, B exchanged

where the interaction Hamiltonian has been assumed to reduce to an effective sum of one-electron terms

$$H^{\text{int}} = \sum_i h^{\text{int}}(i) \qquad (6\text{-}9)$$

The consequences of the important expression (6-8) are studied in Section 6.2.

In the case of a unimolecular reaction, with a perturbation Hamiltonian (6-2b), the perturbation energy is

$$E = E' + E'' = \langle 0|\Delta V_{nn} + \Delta V_{ne}|0\rangle - \sum_{\substack{\text{Excited} \\ \text{states}^*}} \frac{\langle 0|\Delta V_{ne}|*\rangle^2}{E_* - E_0} \qquad (6\text{-}10)$$

where the subscripts 0 and * refer respectively to ground and excited states of the *unperturbed* molecule. The "first-order" perturbation energy has contributions from both nuclear-nuclear and nuclear-electronic potentials, whereas the "second-order" term has contributions from the latter alone. Generally the first term is positive since it represents the change in energy due to the deformation using the fixed starting wave function. Since the molecule starts off at equilibrium, the energy must climb. The second term is negative and represents (in the perturbation formalism) the energy lowering due to the rearrangement or "relaxation" of the wave function as it adjusts to the moving nuclei. This term will be particularly important since it tells us which motions of the molecule are "soft," with a relatively low energy curvature.

6.2 Frontier Orbital Method*

In 1952, Fukui et al.[8-11] proposed a new method for predicting the position of electrophilic substitution in aromatic molecules, based on the relative density of the "frontier" electrons which occupy the highest occupied molecular orbital (HOMO). Similarly, for nucleophilic aromatic substitution, the density of the lowest unoccupied orbital (LUMO) was proposed as the determinant factor. The idea was similar to one put forward two decades earlier by Hückel[12] to explain the position of reduction of aromatics by alkali, and its ease, a function of the energy of the LUMO. In 1964 the frontier orbital approach was extended in a crucial paper[13] to bimolecular reactions. There the favorable nature of the Diels-Alder reaction was explained on the basis of favorable *overlap* between HOMO of butadiene and LUMO of ethylene, concurrent with that between HOMO of ethylene and LUMO of butadiene (Fig. 6-1). The HOMO of butadiene, the π_2 second molecular orbital, has a nodal plane cutting through the central CC bond: its amplitudes at the carbon termini are out of phase. This is also true for the lowest vacant (LUMO) orbital of ethylene, the antibonding π^* orbital, which is out of phase between the two carbon atoms. Hence the two orbitals can overlap positively in a remarkable "fit." This is true regardless of the absolute sign chosen for the butadiene orbital: the ethylene molecule can always "choose" the overall sign of its π^* orbital to fit the butadiene orbital. In a similar fashion, the amplitudes of the LUMO of butadiene π_3^* are in phase on the terminal carbon atoms; thus a perfect fit is possible with the HOMO of ethylene, the π orbital. It should also be noticed that orbitals π_2 and π_3^* of butadiene have precisely large amplitudes on the terminal atoms (0.601 vs. 0.372 on the central atoms) so that the situation is truly optimal for strong interaction with ethylene.

The "frontier" orbital interactions in Fig. 6-1 are no less than a pictorial illustration of (6-8). A large second-order *stabilization energy* in the initial stage of the reaction requires large matrix elements

$$\langle \psi_i^A | h^{\text{int}} | \psi_p^B \rangle \tag{6-11}$$

between occupied orbitals of one molecule and vacant orbitals of the other† and

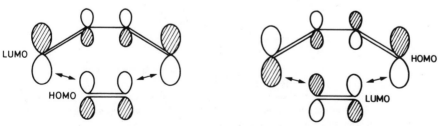

LUMO

HOMO

HOMO

LUMO

FIGURE 6-1.   Favorable frontier orbital interactions in the Diels-Alder reaction. The two interactions occur simultaneously.

*For pertinent information, see Refs. 4–7.

†Generally the overlap term $\langle \psi_i^A | \psi_p^B \rangle E_0$ is small in front of the Hamiltonian term $\langle \psi_i^A | h^{\text{int}} | \psi_p^B \rangle$.

a low excitation energy; hence the optimization of frontier orbital interactions. In a sense these requirements could have been obtained directly by consideration of a traditional molecular orbital interaction diagram (Fig. 6-2). They follow also from theories in which the interaction between two conjugated molecules with overlapping orbitals is described in terms of their separate π molecular orbitals.[14-16] A direct consequence of (6-8) is that, for two different frontier–orbital pair interactions that have the same orbital overlap, the dominant one is that with the smallest energy gap [smallest energy denominator (in 6-8)].

Yet there are two puzzling features that remain to be explained. First, according to (6-8) or to the orbital interaction diagram (Fig. 6-2), the total energy should decrease as the molecules meet; yet it is well known that the total energy rises toward the activation barrier. Second, according to Fig. 6-2, the separation between frontier orbitals in opposing molecules should increase after interaction. Yet it is just the opposite which occurs as shown in Fig. 4-7 for the Diels-Alder reaction.

The answer to these two puzzles lies in the *first-order* energy E', which contains (1) the repulsion energy due to the overlap between the filled orbitals of both systems (compare with He—He, Section 4.1) and (2) the distortion energy due to the nuclear displacements as the molecules rearrange during their approach. The repulsion energy overrides the second-order stabilization energy and the total energy increases.* At the orbital interaction level, the HOMO of ethylene is pushed up by its interaction with π_1 of butadiene which has the same symmetry (recall Fig. 4-5); also the LUMO of ethylene is pushed down by π_4^* of butadiene (Fig. 6-3) but this gives no contribution to E'. To this must be added the actual effects of the nuclear deformations (double-bond stretching for all three double bonds) and of the charge transfer from the HOMOs to the LUMOs: automatically the two HOMOs, bonding in the double bond regions, rise—while the two LUMOs, antibonding there, descend. These results follow Fukui's "principle of narrowing of inter-frontier level separation".[4]

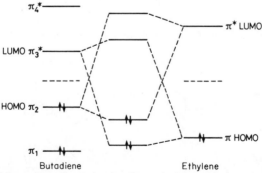

FIGURE 6-2.   Orbital interaction scheme showing the required favorable second-order orbital interactions for overall stabilization in the Diels-Alder reaction.

*The relative values of σ framework distortion energy and stabilizing π interactions in aromatic and homoaromatic systems have been studied by Houk et al.[17]

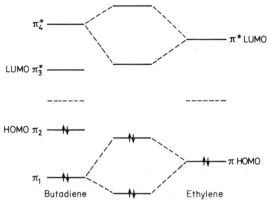

FIGURE 6-3.   Orbital interaction scheme showing the first-order orbital interactions in the Diels-Alder reaction.

If the repulsion energy dominates the *quantitative* chemical behavior, *qualitatively* the stabilization energy due to frontier interactions is much more sensitive to the detailed features of the electronic ornamentation surrounding the two molecules. Exclusion repulsion occurs between two electronic shells whatever the nooks and crannies of the amplitude on specific atoms or bonds. Frontier interactions, on the contrary, involve a specific fit between a hole on one side and an electronic periphery on the other (like fingertips in a glove); they are extremely sensitive to the geometrical details of these on both molecules. This is at the origin of the success of Fukui's frontier theory in predicting the preferred outcome between competing and apparently balanced pathways, such as in the regioselectivity of cycloadditions (see Section 6.3).

6.3   Reactions Controlled by Frontier Orbital Interactions

The frontier orbital method has been remarkably successful in explaining the pathways of widely different types of reaction. Examples are given in the following subsections.

Electrocyclic Ring Closure of Polyenes (See Also Section 5.3)

Curiously this example, which did much to give frontier theory widespread fame, concerns a unimolecular reaction. In their very first paper, Woodward and Hoffmann[18] showed how the stereochemical outcome of the ring closure of butadiene to cyclobutene is predictable from consideration of the HOMO of butadiene (π_2). They required that the frontier orbital, which must overlap with itself in the ring closure, do so *in phase*. The relative phases of the terminal atomic orbitals ϕ_1 and ϕ_4 (Fig. 6-4) are predetermined by conjugation in the diene. For positive overlap to occur, a specific mode of ring closure, the "conrotatory" mode must be obtained. Here the frontier method is used in its original manner,[8-12] with the assumption that the total energy behavior parallels that of a single orbital. Identical predictions were later obtained by using correlation diagrams.[19]

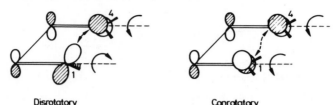

Disrotatory   Conrotatory

FIGURE 6-4.   Conrotatory and disrotatory ring closure in butadiene (molecular orbital π_2 shown).[18]

An interesting facet of such a frontier-controlled reaction is that the conrotatory closure requires a passage through zero overlap between ϕ_1 and ϕ_4. Initially negative, the overlap becomes positive in the product. Hence, all other things being equal, a conrotatory "symmetry-allowed" reaction should be slightly less facile[20] than a disrotatory "symmetry-allowed" one—a barrier appearing in the energy of the HOMO when $S_{14} = 0$.[20,21]

The S_N2 Reaction[22-24]

Here frontier theory is applied to the bimolecular interaction between an attracting nucleophile and a substrate with a tetrahedral carbon atom, one bond of which is to be broken:

$$Y^- + R_3CX \rightarrow R_3CY + X^- \tag{6-12}$$

The attacking nucleophilic reagent generally has a high occupied orbital (a p lone pair for Cl^- or OH^-), so we seek the best possible interaction between this orbital and the LUMO of substrate—the σ^*_{CX} antibonding orbital on the CX bond. Indeed, X is electronegative (F, Cl, etc.) and its bond orbitals are lower in energy than those of the CR bonds.

The two possible interaction schemes—inversion or retention—are illustrated in Fig. 6-5. The electronegative substituent bears a large amplitude of bonding σ_{CX} orbital character, and in the antibonding orbital σ^*_{CX} the amplitude on the carbon atom exceeds that on the halogen. To seek out this large amplitude the reagent Y can attack on the frontside (leading to retention of configuration), where it overlaps with the bigger of the two lobes of carbon. But it then finds itself having out-of-phase overlap with the orbital of the leaving group (Fig. 6-5). The other possibility is rear-side attack with inversion of configuration, with a smaller but still significant overlap, and that generally prevails. Of course, as the CX bond stretches, σ^*_{CX} decreases in energy and, concurrent with the interfrontier level narrowing, the interaction with the HOMO of the nucleophile increases and pushes the reaction on.

A direct consequence of this interpretation is that appropriate modifications of the substrate should favor retention. In particular, an increase in the electronegativity difference between departing group X and reaction center should decrease the amplitude of orbital X in σ^*_{CX} and hence lessen the out-of-phase overlap in the approach with retention.[25] Thus substitution at silicon is predicted[25] to give more

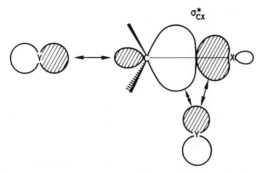

FIGURE 6-5.   Frontier orbital interactions for nucleophilic substitution with inversion and retention.

retention than that at carbon (all other things being equal)—as will also a change in leaving group from chlorine to fluorine or from SR to OR. These results are in agreement with the relatively scarce experimental evidence.[26–28] In a similar vein, retention may be expected if the attacking reagent Y carries a filled d orbital that can overlap appropriately with the frontside part of σ^*_{CX}: this is the main driving force for oxidative addition.

1,3 Dipolar Cycloadditions[6,29,30]

Whether of the azomethine-ylide type

$$\underset{/}{\overset{\backslash}{C}} = \underset{\underset{R}{|}}{\overset{\oplus}{N}} \overset{\ominus}{\underset{\backslash}{\ddot{C}}} \longleftrightarrow \underset{/}{\overset{\backslash}{C}} {=} N {-} \underset{\underset{R}{|}}{\overset{\cdot}{\ddot{C}}} \left(\longleftrightarrow \overset{\backslash}{\underset{/}{C}} {-} \underset{\underset{R}{|}}{\overset{\cdot\cdot}{\ddot{N}}} {-} \overset{\cdot}{\underset{\backslash}{C}} \right) \qquad (6\text{-}13)$$

or of the nitrile-ylide type

$$-C{\equiv}\overset{\oplus}{N}{-}\overset{\ominus}{\underset{\backslash}{\ddot{C}}} \longleftrightarrow -\overset{\cdot}{C}{=}N{-}\overset{\cdot}{\underset{\backslash}{C}} \qquad (6\text{-}14)$$

these molecules, coined "1,3 dipoles" by Huisgen (to emphasize their 1,3 additions), although from the point of view of charge location they are really 1,2 dipoles, are characterized by a four-electron π system similar to that of the allyl anion. The nitrile-ylides have an additional pair of "in-plane π'-type" electrons. The favorable frontier orbital interactions are shown in Fig. 6-6. The HOMO of the 1,3 dipole, similar to the nonbonding orbital of the allyl skeleton, is antisymmetric with respect to the plane of symmetry preserved in the addition geometry. It overlaps strongly with the LUMO of olefin. The opposite interaction—LUMO of dipole with HOMO π of ethylene—should be weaker because of the larger energy gap. However appropriate substitution on the olefin can modify this state of affairs.

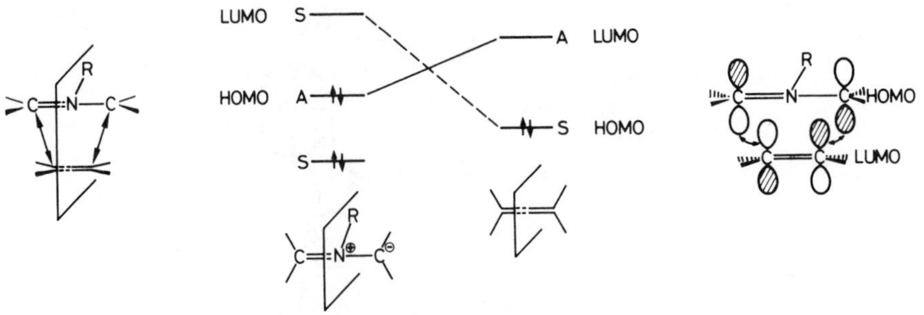

FIGURE 6-6.   Frontier orbital interaction scheme in 1,3 dipolar cycloadditions.

Accordingly, Houk[6] and collaborators have shown that for electron-rich "dipolarophiles" such as styrene or enol ethers, the interaction of 1,3 dipole LUMO with alkene HOMO becomes the dominant frontier interaction. Such a change in predominant interaction can bring about a reversal in "regioselectivity," with preferred formation of head-to-head or head-to-tail adducts. For instance, in the case of nitrones with good donor substituents[7]

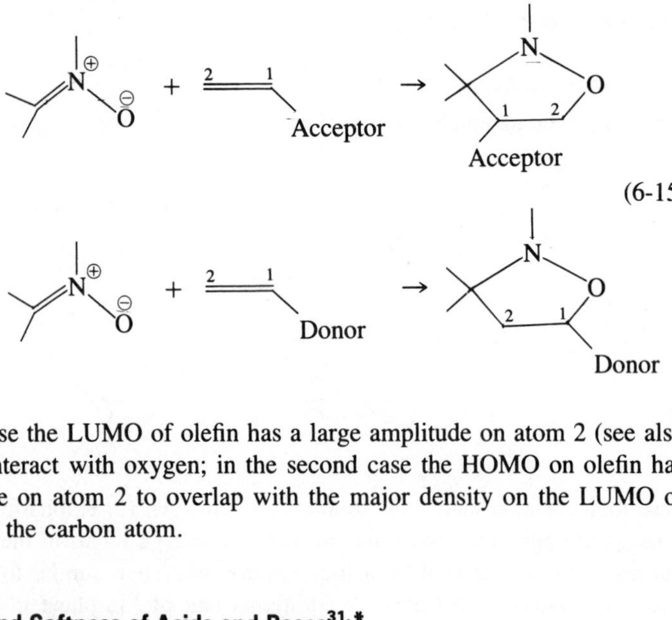

$$(6\text{-}15)$$

In the first case the LUMO of olefin has a large amplitude on atom 2 (see also Fig. 6-12b) to interact with oxygen; in the second case the HOMO on olefin has a large amplitude on atom 2 to overlap with the major density on the LUMO of the 1,3 dipole at the carbon atom.

6.4   Hardness and Softness of Acids and Bases[31, *]

Thus far we have emphasized the interaction between neutral systems. What happens if the interacting molecules carry net charges, as in acid-base neutralization? Clearly, the frontier-orbital controlled energy adds to the Coulombic energy. For two acid-base pairs with the same mutual electrostatic energy, the frontier energy

*See also Ref. 5 for a review.

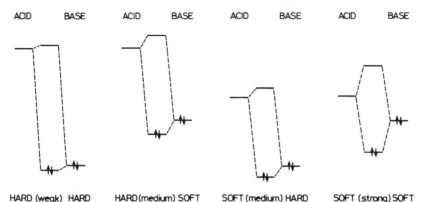

FIGURE 6-7. The four possible cases for the frontier interactions in acid-base neutralization (LUMO of acid, HOMO of base). Stabilization energy in parentheses. The scheme is valid only in protic solution (see p. 169).

will be a modulating or discriminating effect. Figure 6-7 shows four possible cases depending on whether the LUMO of acid is high or low and whether the HOMO of base is high or low. The conclusions drawn from Fig. 6-7 remain valid even for Lewis acids and bases since the frontier interactions subsist whether a net charge is present or not.

From Fig. 6-7 the stabilization due to the frontier interactions is largest for an acid with low-energy LUMO (metals with half-empty s orbital such as Cu^+, Ag^+, Hg^{2+} or neutral electron acceptors such as Br_2, tetracyanoethylene) and a base with high-energy HOMO (CN^-, RS^-, neutral phosphine R_3P, carbonyl CO groups). The acids and bases that we have just cited correspond exactly to two categories that Pearson[31] established and called "soft" acids and "soft" bases. Pearson made the crucial point that soft acids react faster and form stronger bonds with soft bases, whereas "hard" acids (the proton H^+, alkali cations Li^+, Na^+, K^+ or neutrals such as BF_3, $AlCl_3$) react better with "hard" bases (hydroxyl OH^-, the halogen anions fluoride F^- or chloride Cl^-; sulfate SO_4^{2-}, nitrate NO_3^-, but also neutral ammonia NH_3 or amines RNH_2). Klopman[32] later showed how this observation is related to frontier interactions. The strong preference of soft for soft is a consequence of the large HOMO-LUMO interaction, as shown previously. What is less clear is the origin of the strong preference of hard for hard: in Fig. 6-7 the orbital interactions alone would favor "hard" $\leftrightarrow$ "soft" over "hard" $\leftrightarrow$ hard"; OH^- should react better with Cu^+ than with Na^+, contrary to experiment.

Klopman explains the "hard" $\leftrightarrow$ "hard" preference by the electrostatic energy term. Generally, if a base has a low–lying HOMO electron pair (hard), this pair must be tightly held, implying a strong electrostatic field and a small polarizability. (By "polarizability" we mean the extent to which the electron clouds of the molecule can distort under the influence of an external field, such as caused by an approaching ion. A facile distortion requires the presence of *both* HOMOs and LUMOs, which can be easily mixed by the oncoming perturbation). Similarly, the presence in an acid of a high-energy first vacant orbital indicates again a weak polarizability, a tight shell of electrons, and a strongly electropositive molecule, all typical of a hard acid. So both "hard" reactants are characterized by a small radius, low polarizability, and strong electrostatic field. A large stabilization then arises from the encounter of a hard acid and a hard base (Fig. 6-8), more than compensating for their weak frontier-orbital stabilization. As soon as a "soft" component is involved, the diffuse nature of its electron shell decreases the electrostatic attraction with the counterion. Indeed, the full $1/R$ attractive energy is reached only when the ionic spheres do not interpenetrate.* Another explanation would assume, as Pearson did, that the net charge is actually larger on the hard systems, from which their mutual preference follows immediately.

Hence the Pearson concept really covers two different types of reaction: "frontier-controlled" and "charged-controlled."[15,32] Simple expressions exist [4-7,14-16] for the leading terms in the energy of such reactions, for pairs of interacting atoms r (on base) and s (on acid), respectively. The frontier term is [see (6-8)]

$$E_{\text{frontier}} = -2 \frac{(\sum_r \sum_s c_r^{\text{HOMO}} c_s^{\text{LUMO}} \beta_{rs})^2}{E_{\text{LUMO}} - E_{\text{HOMO}}} \quad \begin{array}{l} \text{(assuming a single} \\ \text{HOMO-LUMO} \\ \text{interaction to predominate)} \end{array} \quad (6\text{-}16)$$

where β_{rs} is the matrix element of h^{int} (6-9) between atomic orbitals on r and s and c_r^j is the coefficient of molecular orbital j on atom r. The electrostatic term contains the Coulomb attraction appropriately modulated by the dielectric constant of the medium[33], to which must be added a "desolvation energy" as the initially solvent-separated ions come together.

The role of the solvent is actually more important than appears at first sight in determining the properties of hard and soft acids and bases. The previous statement to the effect that a high LUMO implies a strongly electropositive ion (Li^+) needs elaboration: in fact, a calculation on the isolated Li^+ ion shows the empty $2s$ orbital to be much lower in energy than that of larger and supposedly softer ions. Similarly, the HOMO of small anions (OH^-, F^-) are quite high in the gas phase, as would

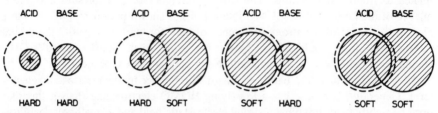

FIGURE 6-8.   Electrostatic interactions: the four possible cases for acid-base neutralization. The LUMO of the acid is shown in dotted lines.

*According to Gauss' theorem, a fraction of negative charge that is not outside the positive sphere will not feel the full attraction of that sphere.

be expected from the large interelectronic repulsion generated for electrons concentrated in a small volume.[34] So these ions taken alone have the orbital characteristics that we ascribed to soft systems! It is only *solvation,* by protic solvents, which gives *hard* characteristics to small ions.[34] For the small anions, the strong positive solvation shell pulls down the HOMO. For the small cations, the negative counterions push the LUMO back up. The small ions also acquire part of their hard character at the beginning of the acid-base reaction, by direct interaction with the substrate, whose Coulombic field has a similar stabilizing effect as protic solvents do. On the other hand, complexation of a Li^+ cation by crown ethers or cryptands does make it behave in a "soft" manner, with frontier orbital control due to the low LUMO of the ion that reacts as if it were isolated[35] and also as if it were larger than its actual size.

6.5 Frontier Interactions for Photochemical Reactions

Frontier orbital theory deals equally well with reactions when one of the partner molecules is excited. The major change in the interaction energy comes from the first-order energy E' (6-5a). Since excitation creates a "hole" in the closed shell of one reaction partner, the exclusion repulsion due to the Pauli principle (Section 4.1) decreases. At the same time a first-order stabilization energy arises from the mixing between the orbital of the excited electron and the nearest empty orbital in the partner. The two effects are illustrated in Fig. 6-9.

In such reactions the first-order perturbation energy thus generally swamps the second-order effects and is the predominant factor in controlling photochemical reaction paths. This is true regardless of whether the reaction occurs in a singlet or triplet excited state. Again frontier theory concurs with the correlation diagram method, for instance, in predicting the ease of photocyclization of two olefins in

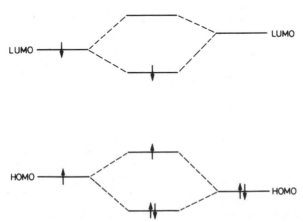

FIGURE 6-9. Interaction between an excited molecule and a ground-state molecule (frontier orbitals) showing the first-order stabilization of the excited electron by mixing of its orbital with that of its partner.

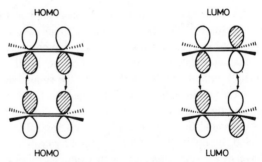

FIGURE 6-10.   Favorable frontier orbital interactions in the face-to-face photodimerization of ethylene.

a face-to-face manner (Fig. 6-10). The empty orbitals overlap in phase, as do the filled orbitals. Analytically, the frontier interaction energy is now dominated by the term[16]

$$E_{\text{frontier}} = -(|\sum_r \sum_s c_r^{\text{LUMO}} c_s^{\text{LUMO}} \beta_{rs}| + |\sum_r \sum_s c_r^{\text{HOMO}} c_s^{\text{HOMO}} \beta_{rs}|) \quad (6\text{-}17)$$

to which must be added a second-order term equal to (6-16), but divided by 2 to account for the loss of a HOMO-LUMO interaction. Equation (6-17) can readily be used[16] to compare the various possible incipient pathways for reactions such as butadiene photodimerization or the photodimerization of tropone.

In summary, favorable paths for nonionic bimolecular reactions are determined by maximizing the following orbital interactions:

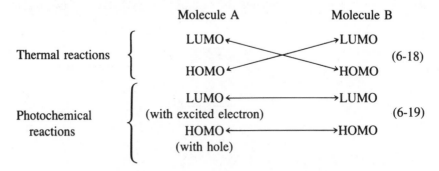

Frontier orbital theory has also been applied to free-radical reactions, possibly with a somewhat less persuasive success. The reaction outcome is considered to depend on the interaction between singly occupied orbital (SOMO) of radical and either HOMO or LUMO of substrate. The different possibilities are illustrated in Fig. 6-11.[5] Successful, for instance, in explaining alternating copolymerization,[5] the method fails sometimes by not being able to choose *a priori* which of the two interactions (SOMO-HOMO or SOMO-LUMO) predominates.[36]

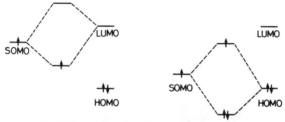

FIGURE 6-11.   Important frontier interactions for *(left)* nucleophilic radical with high-energy odd-electron orbital (SOMO); *(right)* electrophilic radical with low-energy odd-electron orbital.[5] From *Frontier Orbitals and Organic Chemical Reactions* by I. Fleming. Copyright 1976 by Wiley & Sons, Ltd. Reprinted by permission of the publishers.

6.6   Other Types of Orbital Interaction: Donor-Acceptor Interactions and Subjacent Orbital Control

We have already examined (Section 4.9) cases where there is partial breakdown of orbital symmetry control, with forbidden reactions becoming feasible. The frontier orbital method, or more generally, the perturbation approach derived in Section 6.1, accounts equally well for these phenomena.

Donor-Acceptor Interactions[37]

Let us return to the thermal cycloaddition of two olefins: as shown in Fig. 6-12*a*, the orbital-symmetry-forbidden character of the reaction is expressed here by the absence of any favorable HOMO-LUMO interactions. In the presence of a donor substituent (propylene CH_3—CH=CH_2, etc.) or an acceptor substituent (cyanoethylene CN—CH=CH_2, etc.) the situation is strongly modified. An olefin substituted by a donor that brings a pair of electrons into conjugation with the π bond becomes isoelectronic with the allyl anion[5]:

$$(D = \text{—}CH_3, OCH_3, \text{—}NH_2, \text{etc.}) \qquad \qquad \sim \qquad \qquad (6\text{-}20)$$

If it is substituted by an acceptor, which generally adds a π bond with a low-lying π^* orbital to the system, the olefin molecule becomes isoelectronic with butadiene:

$$(A = \text{—}CN, \text{—}NO_2, \text{etc.}) \qquad \qquad \sim \qquad \qquad (6\text{-}21)$$

The amplitudes of the frontier orbitals follow immediately from these considerations. They are shown qualitatively in Fig. 6-12*b* and quantitatively[6] for the olefin part of electron-rich and electron-poor alkenes in Fig. 6-13 [note that the LUMO of donor-substituted olefins does not obey too well the amplitude expected from (6-20), which predicts a larger amplitude on the substituted atom].

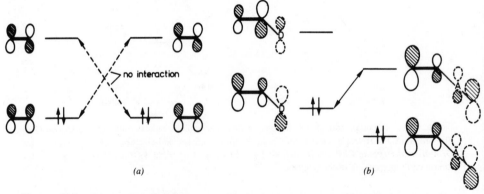

FIGURE 6-12.   Frontier orbital interactions for thermal olefin face-to-face dimerization: *(a)* nonpolar case; *(b)* donor-acceptor case. Orbitals on donor and acceptor substituents shown in dotted lines.

Inspection of Fig. 6-12*b* shows that substitution brings about two major changes in the interaction between frontier orbitals: (1) the unbalanced amplitudes on HOMO of electron-rich olefin and LUMO of electron-poor olefin now make for a significant overlap between the two terminal atoms (2,2′) in a face-on "head-to-head" approach; this approach is the same that would be predicted from simple polarity considerations:

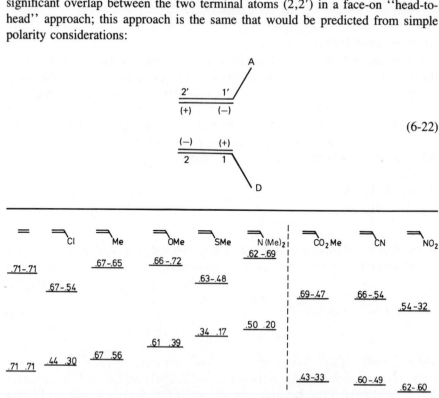

$$(6-22)$$

FIGURE 6-13.   Frontier orbitals of electron-rich and electron-deficient alkenes. Relative energies also shown.[6] Reprinted with permission from K. N. Houk, p. 181 of *Pericyclic Reactions*, Vol. 2, edited by A. P. Marchand and R. E. Lehr. Copyright 1977 by Academic Press.

(2) the HOMO on donor side and LUMO on acceptor side are very close in energy. All the conditions are thus present for a facile reaction to occur, most likely in two steps through a zwitterion. As the reagents interact, charge transfer occurs from HOMO to LUMO, so that the charge-transfer configuration (or resonance structure, depending on the model used) gives a significant contribution to the transition state. Hence this method is capable of reproducing qualitatively the "spectral" behavior from polar to nonpolar cycloadditions[37] and from concertedness to nonconcertedness.

Subjacent Orbital Control[38–40]

Orbital interaction schemes sometimes reveal features that do not show up clearly in correlation diagrams. This is particularly true of certain "secondary" interactions. For example, symmetry-allowed sigmatropic reactions such as (6-23) are well interpreted by the "primary" interaction between the atomic orbital of the migrating group and the nonbonding orbital on the remaining skeleton[41]:

$$\text{(6-23)}$$

If the symmetry-forbidden pathway (retention instead of inversion at the migrating group) is chosen, however, the primary interaction vanishes. There remain secondary interactions between the atomic orbital on migrating group and both filled bonding and empty antibonding orbitals on the other fragment; *both* effects are stabilizing. The overall result is illustrated in Fig. 6-14 for the suprafacial 1,3-sigmatropic rearrangement of 1–butene in which the methyl group is assumed to retain its initial configuration.

In the transition state—formally of the antiaromatic cyclobutadienoid type (Sections 3.9 and 5.4)—the interaction between atomic orbital of CH_3 and bonding π allylic orbital destabilizes one electron in the former and stabilizes two electrons in the latter. The interaction between the π^* allylic orbital and the atomic orbital

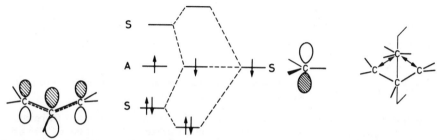

FIGURE 6-14. Subjacent orbital control (symmetries relative to vertical plane). The atomic orbital of the migrating group interacts favorably with the antibonding and bonding orbitals of the allyl skeleton, which share the same (S) symmetry.

restabilizes the odd electron in a compensating manner. Overall, the atomic orbital of CH_3 stays put in energy whereas there occurs a net stabilization of a bonding electron pair. This effect, coined "subjacent orbital control,"[38–40] is apparently sufficient to make the Woodward-Hoffmann symmetry-forbidden pathway predominate over a stereochemically random diradical pathway in those reactants where the symmetry-allowed path is sterically hindered. The bicyclo [3.2.0]heptene to bicyclo [2.2.1]heptene rearrangement is a case in point.[42]

6.7 Relaxability: Theory[43–48]

We now return to the process of a single molecule undergoing a unimolecular reaction. Can we guess at the reaction pathway occurring after the molecule has received sufficient thermal energy by collisons with other sister molecules? A positive answer is given by the "relaxability" method, which considers the initial distortion of the molecule as a perturbation. Then the energy change of the molecule can be evaluated as in all perturbational treatments. One must know all the states, ground and excited (including continuum states) of the starting molecule as well as the coupling, through the perturbation, between ground state and excited states. In other words, the detailed surface of the energy "bowl" *surrounding* the equilibrium geometry—an infinity of points—is known through a set of infinite wave functions *at* that geometry. In principle, the number of unknowns is the same. In practice, certain low-lying excited states may give sufficiently overwhelming contributions to the perturbation energy so as to indicate directly the easiest pathway out of the bottom of the bowl (Fig. 6-15). In particular, if there is a very low–lying state that couples strongly with ground state, the molecule will need little energy to distort; one speaks of a "pseudo–Jahn-Teller" effect.*

The appropriate starting expression is

$$E = \langle 0|\Delta V_{nn} + \Delta V_{ne}|0\rangle - \sum_* \frac{\langle 0|\Delta V_{ne}|*\rangle^2}{E_* - E_0} \tag{6-10}$$

excited states

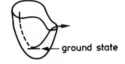
ground state

FIGURE 6-15. Excited states of the starting molecule can be used to find the best pathway out of the bowl.

*The "pseudo–Jahn-Teller effect" (distortions in systems where two electronic states have their energies not exactly equal but only close to each other), was first described by Öpik and Pryce.[49] The term "pseudo–Jahn-Teller effect" itself seems to have originated in the Cambridge University Theoretical Chemistry Laboratory in the late 1950s or early 1960s (e.g., see den Boer et al.[50]). There are numerous examples of inorganic complexes in which distortions are promoted by pseudo Jahn-Teller effects[51].

It is convenient to expand the potential-energy operator changes ΔV along the coordinate Q along which we are searching. Then, to second order in the nuclear displacements, the expression

$$\frac{\partial^2 E}{\partial Q^2} = \langle 0 \left| \frac{\partial^2 (V_{nn} + V_{ne})}{\partial Q^2} \right| 0 \rangle - \sum_* \frac{\langle 0 \left| \frac{\partial V_{ne}}{\partial Q} \right| * \rangle^2}{E_* - E_0} \tag{6-24}$$

gives the curvature of the electronic energy relative to the coordinate Q, that is, twice the force constant for motion along Q. Bader[43] was the first to use this expression to compare the force constants of different vibrational normal modes in simple polyatomic molecules. In the famous example of CO_2, the lowest excited state has $^1\Sigma_u^+$ symmetry, the ground state having $^1\Sigma_g^+$ symmetry. The second term of (6-24) thus gives a large negative contribution to the perturbation energy for a $\partial V_{ne}/\partial Q$ operator with Σ_u symmetry. But $\partial V_{ne}/\partial Q$ has the same symmetry as Q, so the Σ_u^+ "antisymmetric" normal mode of nuclear motion

$$O\rightarrow \leftarrow C=O\rightarrow \tag{6-25}$$

has a large negative stabilization term (the total perturbation energy remains positive, however, because of the first term). Such a stabilizing contribution is absent for the symmetric stretching motion

$$\leftarrow O=C=O\rightarrow \tag{6-26}$$

Hence the normal mode represented by (6-25) has a lower force constant than that shown in (6-26). Alternatively, we can say that the "interaction constant" k_{12} defined in the expansion of the molecular energy

$$E = \frac{1}{2} k(\Delta r_1^2 + \Delta r_2^2) + k_{12} \Delta r_1 \Delta r_2 \tag{6-27}$$

is positive.

Physically, the first term of (6-24) represents the energy change due to the nuclear motion within a fixed electronic density (a sort of "classical" force constant). The second term represents the decrease in energy due to the rearrangement of the electron clouds as the nuclei are displaced. It can be called the "relaxability"[44,22] of the molecule along the coordinate Q. Let us now more carefully study the matrix elements

$$\langle 0 \left| \frac{\partial V_{ne}}{\partial Q} \right| * \rangle \tag{6-28}$$

The size of such matrix elements will depend on (1) the relative symmetry of ground 0 and excited * states (the element vanishes unless the direct product of the representations of the two functions contains the representation of the displacement Q), and (2) the location of the overlap between the wave functions Ψ^0 and $\Psi*$; more specifically, the "transition density"[52] ρ_0* between Ψ^0 and $\Psi*$ must be located in the region of the nuclei which contribute to the diplacement Q. (For a closed-shell ground-state Ψ^0 and an excited-state Ψ^* that differ by a one-electron transition between two orbitals ψ_i and ψ_j, ρ_0* is simply the product $\sqrt{2}\psi_i\psi_j$ of the two orbital amplitudes). The contribution of an excited state to the relaxability toward a given mode Q depends on these two factors and, of course, on the size of the excitation energy.

However, the comparison between two competing modes Q_1 and Q_2 depends on more than just the relaxability toward these modes. The classical force constant plays a crucial role: the first term of (6-24) is, for instance, much smaller for a bending motion than for a stretching motion; hence a bending motion is almost always preferred over a stretching motion, regardless of the relative relaxabilities! Valid comparisons must be restricted to modes involving the same set of internal valency coordinates (a group of symmetry-equivalent bond length changes or bond angle changes). Only then is it possible to predict lowest-energy pathways from the symmetries of ground and lowest excited states. Even so, a further restriction concerns the location of the transition density ρ_0*, which must be large near the nuclei that contribute to the mode Q. For example, in planar hydrocarbons $\sigma \rightarrow \sigma*$ transitions are much more effective than the lower-lying $\pi \rightarrow \pi*$ transitions in promoting in-plane nuclear motion. Similarly, the $n, \pi*$ transition in acetone can promote out-of-plane motion of the carbonyl carbon, but hardly that of the methyl carbon atoms.

6.8 Relaxability: Applications

We now give several illustrations of the relaxability method as applied to actual reaction paths.

Interconversion of PH_5[53]

The PH_5 molecule, a trigonal bipyramid at equilibrium, needs only 16 kJ/mole to reach the square pyramid geometry.[54] The orbital levels for the two forms, with respective D_{3h} and C_{4v} symmetries, are shown in Fig. 6-16. The best interconversion motion is known* to be the Berry pseudorotation,[56] which brings one D_{3h} geometry into a distinct one—through the C_{4v} geometry—by exchanging two axial and two equatorial ligands.

What does the relaxability method say about the motion from D_{3h} to C_{4v}, or backward? In the equilibrium geometry, the lowest excited state, corresponding to the molecular orbital transition $4a_1' \rightarrow 3e'$, has symmetry e'. There are three e' normal modes of vibration in a trigonal bipyramid.[57] One of these is a PH stretching mode, which has a relatively large "classical" force constant [first term in (6-24)]

---

*For a sound calculation of the potential surface for pseudorotation, see Ref. 55.

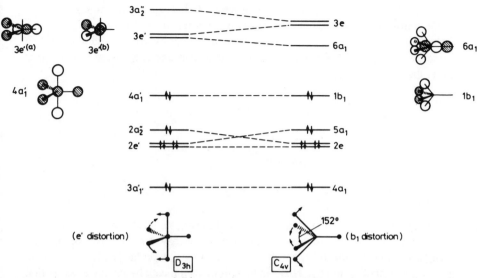

FIGURE 6-16. Orbital energy levels of D_{3h} and C_{4v} structures of PH_5, together with their correlation.[53] Distortions which interconvert the two forms and frontier orbitals are also shown.

and can be left aside. The other two e' modes involve $\widehat{HPH}$ bending respectively for two equatorial ligands (ν_6) and for the two axial ligands (ν_7):

$$(6\text{-}29)$$

$$\nu_6\ (e')\qquad\qquad \nu_7\ (e')$$

A proper combination of these two motions opens the equatorial angle and closes the axial angle, precisely as required by the Berry pseudorotation mechanism. Hence the symmetry of the lowest energy motion is well rationalized by the theory. This is true also for the detailed nuclear motions. Indeed, the transition density $\rho_{0_*} = 4a_1' \times 3e''^{(a)}$, which resembles $3e''^{(a)}$, tends to push the two equatorial ligands in the directions shown by the arrows in ν_6 (see Fig. 6-16, orbital $3e''^{(a)}$, with nuclei to be displaced away from unshaded regions towards shaded regions: the unshaded lobe in the left-hand $\widehat{HPH}$ angle pushes the two hydrogens apart and opens the angle†).

---

†Quantitatively, however, the argument is not entirely convincing because the movement of the P atom induced by ρ_{0_*} tends to close this angle.

Similarly, in the C_{4v} structure the lowest excited state corresponds to a $1b_1 \rightarrow 6a_1$ transition, with symmetry b_1. The two b_1 normal modes for a square pyramid[53] include one PH stretching mode for the four equivalent hydrogens and one bending mode in which the H⌢PH angle between nonadjacent basal hydrogen atoms opens while the other H⌢PH angle closes:

$$(6\text{-}30)$$

This is precisely the motion required to distort the C_{4v} structure into a D_{3h} structure (Fig. 6-16). Now we know *a posteriori* that the energy curvature for this motion is negative, so the relaxability for the b_1 motion in C_{4v} is probably even larger than that for the e' motion in D_{3h}. This can be rationalized nicely by the lower excitation energy in the perturbation expression for the C_{4v} molecule (the difference in SCF orbital energies [see (1-19)] is 0.496 a.u. for $1b_1 \rightarrow 6a_1$ but 0.557 a.u. for $4a'_1 \rightarrow 3e'$).[54]

Coplanar Decomposition of Cyclobutane[58]

We return one last time to the cyclobutane $\rightarrow$ (two ethylenes) reaction (see Sections 4.2 and 5.2). Let us try to use the relaxability method to compare the concerted decomposition of the ring and a stepwise decomposition (Fig. 6-17a). Both motions involve CC stretching; thus it is legitimate to concentrate on the relaxability term.* The four CC stretching *symmetry* coordinates are shown in Fig. 6-17b for the molecule reduced to a square ring. These coordinates already enclose the required information on the symmetry of the two paths—and it is not necessary to draw out the normal coordinates, which would have other nuclear motions (CH stretch) mixed in. It is readily shown[58] that

$$Q_{\text{concerted}} = \frac{1}{\sqrt{2}}(Q_{A_{1g}} + Q_{B_{1g}})$$

$$Q_{\text{two-step}} = \frac{1}{2}(Q_{A_{1g}} + A_{B_{1g}} + \sqrt{2}Q_{E_u^a})$$

$$(6\text{-}31)$$

Although both the concerted and stepwise motions involve a mixture of coordinates, it is permissible to say that a unique symmetry mode characterizes each motion: B_{1g} the concerted motion, E_u^a the stepwise motion (see Figs. 6-17a, b).

Now these CC stretching modes involve exclusively displacements of carbon atoms in the molecular plane; hence the appropriate excitations in (6-24) must

---

*Indeed, a proper normalization of the two motions, such as that done in (6-31), should lead to identical classical force constants.

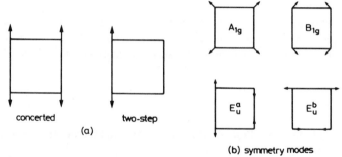

(a)

(b) symmetry modes

FIGURE 6-17. Motions of the carbon skeleton in cyclobutane: *(a)* concerted two-step; *(b)* symmetry modes.

involve transitions between molecular orbitals (σ_{CC} or σ_{CC}^*) built on CC bond orbitals. This ensures that the "transition density" is localized in the regions of nuclear motion. The appropriate molecular orbitals are shown in Fig. 6-18. The lowest excited state has symmetry

$$e_u \rightarrow a_{2g} \equiv e_u \qquad (6\text{-}32)$$

that indicates a preference for two-step motion.

Hence the qualitative information obtained from correlation diagrams is again present here, in the manifold of reactant electronic states. This is not surprising; for instance, we can identify the high energy of the e_u σ^* orbitals and of the corresponding $e_u \rightarrow e_u$ transition required to favor a b_{1g} motion with the high energy of one same orbital (product *SA* orbital) in Fig. 4-3. But this *SA* orbital is precisely responsible for the forbiddenness in the orbital symmetry diagram through its correlation with a low-energy orbital of the ethylene pair. The equivalence—but clearly less strict nature—of the relaxability method with the correlation diagram method is thus qualitatively established.

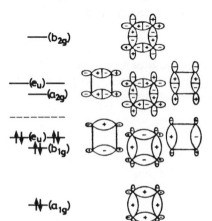

FIGURE 6-18. Molecular orbitals of cyclobutane built out of the CC bond orbitals.[58] Reprinted with permission from J. S. Wright and L. Salem, *J. Am. Chem. Soc.* **91**, 5947 (1969). Copyright 1969 by American Chemical Society.

Transition Structures

It has recently been noted[59] that the classical force constant [first term in (6-24)] should be positive *everywhere* on the multidimensional surfaces for any state. This implies that the negative curvature at transition states is due exclusively to the relaxability term.

6.9   Reaction Paths from Crystal Structures Involving Fragments Identical with the Reaction Center

Bürgi and Dunitz[60,61] have found a truly innovative manner to determine reaction paths. The idea of this method of so-called structural correlation consists in seeking out, for a particular reactive fragment, all the crystal structures in which this fragment is present, either at equilibrium *or distorted* in a manner that resembles the presumed pathway to product. A most surprising feature is that there does exist a sufficient nuumber of X-ray structures in which a distorted form of the reactant can be observed. The geometries of these structures are then plotted together to form an assumed pathway from reactant to product.

   An early example[60] is that of the substitution reaction

$$
X-Cd\overset{S}{\underset{S}{\diagup}}S \;+\; Y \rightarrow X \;+\; S-\overset{S}{\underset{S}{\diagup}}Cd-Y
\tag{6-33}
$$

From the four crystal structures respectively of

$Cd_5 (SCH_2CH_2OH)_4 \cdot SO_4 \cdot H_2O$ (which contains both four-, five- and six-coordinated cadmium atoms)

$Cd (SCH_2CH_2OH)_2$ (with both four- and five-coordinated cadmium atoms)

$Cd (SCH_2CH_2OH)_3I$ (four, five, six coordination)
$Cd [(CH_3CH_2CH_2O)_2PS_2]_2$

Bürgi selects all situations showing a five-coordinated cadmium (or a four-coordinated cadmium with an additional ligand above a triangular face). This gives eleven structures with different pairs of experimental distances x and y:

$$
Y \overset{S}{\underset{\underset{SS}{|}}{\underset{y}{\rule{0pt}{0pt}}\!-\!-\!Cd\!\underset{x}{-\!-}}} X
\tag{6-34}
$$

FIGURE 6-19.   Obtention of reaction *pathway* from a series of *equilibrium* X-ray geometries.[62] Reprinted with permission from K. Müller, *Angew. Chem. Internatl. Ed.* **19**, 1 (1980). Copyright 1980 by Verlag Chemie GMBH.

The axial ligands X and Y are: two iodine atoms (five cases), two sulfur atoms (once), one sulfur atom and one oxygen atom (three cases) and iodine with oxygen (twice).

The fundamental assumption is now made that a *single* potential surface for the required reaction path can be constructed by taking an ensemble of experimental points belonging to *different* molecules (each point really belongs to a different potential surface). These points are all equilibrium geometries from X-ray structures and are thus minima on their respective potential-energy sheet. The relation between the predicted pathway and these minima is illustrated in Fig. 6-19.[62] Of course, there is no guarantee that the assumption is correct. The manner in which crystal forces distort the structure of a fragment into an unusual equilibrium geometry need not correspond to the manner in which the fragment is distorted when it reacts under specific conditions in solution. Yet the success of the method is remarkable—as measured, for instance, by the excellent agreement with sophisticated calculations (e.g., hydride ion addition to carbonyl group).*

Returning to the substitution reaction at cadmium, we can now plot against each other the measured increments Δx and Δy (from normal distances) for the two axial bonds X—Cd and Y—Cd. A smooth curve is obtained (Fig. 6-20). The first half of the curve uses the experimental points, and the other half is obtained by symmetry. The hyperbolic nature of the curve corresponds fairly well to the expectations for a substitution reaction of this type, in which the simultaneous coordination of X and Y must be accompanied by an expansion in Cd—X and Cd—Y bond lengths relative to their normal equilibrium values.

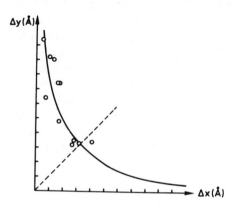

FIGURE 6-20.   Correlation of axial Cd—X distance increment Δx with Cd—Y distance increment Δy in a series of cadmium X-ray structures.[60] Reprinted with permission from H. B. Bürgi, *Inorg. Chem.* **12**, 1321 (1973). Copyright 1973 by American Chemical Society.

*Compare the results in Ref. 61 with those in Ref. 63.

6.10 "Static" Reactivity Indices

An even simpler, but also cruder, approach to chemical reactivity is based on consideration of the "static" properties of the reacting molecule in its ground state: charge, bond order, free valence, polarizability, and so on.[64] These aspects have often been abandoned in the last two decades in favor of more sophisticated methods. Here we concentrate on some more novel aspects of the static approach.

Electrostatic Molecular Potential[65–67]

In Section 6.1 we mentioned the long-standing dream of ascribing to any molecule the information required to understand its reactions with other molecules. Such a dream would become reality if a detailed "reaction field" or "reaction potential" could be defined for a molecule. The incoming partner would be submitted to this potential. Such a potential would obviously have a Coulombic contribution and also an "exchange" part [somewhat similar to (1-20)], which by necessity would depend on the reaction partner. Hence there would have to be a different reaction potential for each partner. The Coulombic potential taken separately should, however, be a property of the molecule alone (the net charges used to determine positions of attack for ionic reagents[64] are crude, "first-order" substitutes for such a potential). Scrocco and his collaborators[65,66] have made a wide use of such molecular potentials.[67] Knowing the wave function for the molecular state, they calculate the function

$$V(r_i) = -\int \frac{\rho(1)}{r_{1i}} d\tau_1 + \sum_{\alpha}^{\text{nucl}} \frac{Z_\alpha}{r_{\alpha i}} \tag{6-35}$$

at each point i in space around the molecule. In (6-35) ρ is the one-electron density and Z_α the charge of nucleus α. The distances of the electronic charge element $d\tau_1$ and of the nucleus α from point i (where V is calculated) are respectively r_{1i} and $r_{\alpha i}$. The potential V is obtained in the form of a map, with contour lines. A deep potential minimum is indicated by concentric contours surrounding a small region. Figure 6-21 shows two typical such maps for cyclopropane (Fig. 6.21a)[65] and formamide (Fig. 6.21b).[67] The first map gives a strong indication that the cyclopropane molecule will edge–protonate* (the map in the plane perpendicular to the molecule has no comparable minimum). The second map reveals beautifully that, in regard to densities rather than orbitals, Pauling's concept of hybrids corresponds to reality: there is a potential energy trough corresponding to each sp^2 lone pair on oxygen. Maps for excited states have also been drawn out and agree well with the known electrophilic or nucleophilic properties of these states.[67] An important problem is the extent of transferability, from one molecule to another, of the potential due to a given group.[66] Apparently, the nature of the neighboring group is not negligible.

---

*For experimental studies of protonated cyclopropanes, see Refs. 68 and 69.

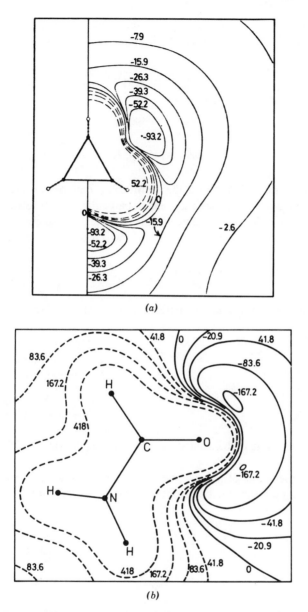

(a)

(b)

FIGURE 6-21.   Electrostatic potential energy maps: (a) cyclopropane, in the ring plane[65]; (b) formamide, in–plane.[67] Attractive potentials are negative. Energies in kJ/mole. Reprinted with permission (a) from R. Bonaccorsi, E. Scrocco and J. Tomasi, *J. Chem. Phys.* **52**, 5270 (1980). Copyright 1970 by American Institute of Physics; (b) from J. Tomasi, p. 191 of *Quantum Theory of Chemical Reactions*, Vol. I, edited by R. Daudel, A. Pullman, L. Salem and A. Veillard. Copyright 1980 by D. Reidel Publishing Company.

Relative Excitation Energies and Different pK Values for Singlets and Triplets[70]
For the equilibrium between a Brönsted acid and its conjugate base

$$A \underset{}{\overset{K_a}{\rightleftarrows}} B + H^+ \tag{6-36}$$

the pK_a value

$$pK_a = - \log K_a \tag{6-37}$$

measures the acidity of A. Since the equilibrium depends on the relative free energies of acid and of conjugate base, and in particular on their relative enthalpies, any process that modifies the relative energies of A and B also changes the acidity. Such is the case of electronic excitation, after which the equilibrium becomes

$$A^* \underset{}{\overset{K_a^*}{\rightleftarrows}} B^* + H^+ \tag{6-38}$$

If the acid has a higher excitation energy than the conjugate base:

$$hv_A > hv_B \tag{6-39}$$

it becomes more acidic in the excited state since the equilibrium (6-38) is displaced to the right.

A typical case is that of phenols, which are more acidic in the excited state. A simple interpretation uses the charge-transfer resonance structure

$$\tag{6-40}$$

that contributes significantly to the lowest excited state. Proton dissociation from this structure that carries a positive charge near the departing hydrogen atom, should be easier than from the ground state described as

$$\tag{6-41}$$

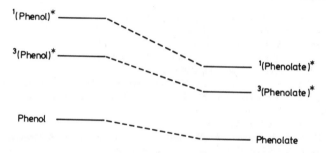

FIGURE 6-22. Qualitative ordering of states in phenols and phenolate anions.

A further distinction can be made between pK_a of excited singlet and that of excited triplet. The change in pK_a from one to the other depends on the *differential* singlet-triplet excitation energy of acid compared with the *differential* excitation energy of conjugated base. For phenols, the decrease in pK from ground state is more marked for excited singlet than for excited triplet ($pK_G = 9.5$, $pK_T = 8.3$, and $pK_S = 3.1$ in 2-naphthol).[71] This implies a progressive tightening (with lower energy) of the state energy levels in the phenolate anion as compared with the neutral parent molecule (Fig. 6-22). The lesser acidity of excited triplet phenol versus excited singlet phenol can be ascribed to the impossibility of excited resonance structures such as (6-40). Relative to the ground state, however, structures such as

$$(6\text{-}42)$$

may be responsible for the slight increase in acidity that is nevertheless observed in the triplet state.

References

1. B. Claverie in *Intermolecular Interactions: From Diatomics to Biopolymers*, edited by B. Pullman (Wiley, New York, 1978), p. 69.
2. B. Claverie, private communication to the author.
3. J. N. Murrell, M. Randic, and D. R. Williams, *Proc. Roy. Soc.* **A284**, 566 (1965).
4. K. Fukui, *Topics Curr. Chem.* **15**, 1 (1970).
5. I. Fleming, *Frontier Orbitals and Organic Chemical Reactions* (Wiley, New York, 1976).
6. K. N. Houk in *Pericyclic Reactions*, Vol. 2, edited by A. P. Marchand and R. E. Lehr (Academic, New York, 1977), p. 181; K. N. Houk, *Topics Curr. Chem.* **79**, 1 (1979).
7. J. Sims and K. N. Houk, *J. Am. Chem. Soc.* **95**, 5798 (1973).
8. K. Fukui, T. Yonezawa, and H. Shingu, *J. Chem. Phys.* **20**, 722 (1952).
9. K. Fukui, C. Nagata, and H. Shingu, *J. Chem. Phys.* **22**, 1433 (1954).

10. K. Fukui, T. Yonezawa, and C. Nagata, *Bull. Chem. Soc. Jap.* **27**, 423 (1954).
11. K. Fukui, T. Yonezawa, and C. Nagata, *J. Chem. Phys.* **31**, 550 (1959) (see discussion).
12. E. Hückel, *Z. Physik* **76**, 628 (1932).
13. K. Fukui in *Molecular Orbitals in Chemistry, Physics and Biology,* edited by P. O. Löwdin and B. Pullman (Academic, New York, 1964), p. 513.
14. M. J. S. Dewar, *J. Am. Chem. Soc.* **74**, 3341, 3345, 3350, 3353, 3357 (1952).
15. G. Klopman, *J. Am. Chem. Soc.* **90**, 223 (1968); G. Klopman and R. F. Hudson, *Theoret. Chim. Acta (Berl.)* **8**, 165 (1967).
16. L. Salem, *J. Am. Chem. Soc.* **90**, 543,553 (1968).
17. K. N. Houk, R. W. Gandour, R. W. Strozier, N. G. Rondan, and L. A. Paquette, *J. Am. Chem. Soc.* **101**, 6797 (1979); W. L. Jorgensen, ibid. **97**, 3082 (1975).
18. R. B. Woodward and R. Hoffmann, *J. Am. Chem. Soc.* **87**, 395 (1965).
19. R. B. Woodward and R. Hoffmann, *J. Am. Chem. Soc.* **87**, 2045 (1965).
20. L. Salem, unpublished results.
21. D. T. Clark and D. R. Armstrong, *Theoret. Chim. Acta (Berl.)* **14**, 370 (1969).
22. L. Salem, *Chem. Br.* **5**, 449 (1969).
23. K. Fukui, *Bull. Chem. Soc. Jap.* **38**, 1749 (1965).
24. R. G. Pearson, *Chem. Eng. News* **48**, 66 (1970).
25. N. T. Anh and C. Minot, *J. Am. Chem. Soc.* **102**, 103 (1980).
26. L. H. Sommer, W. D. Korte, and P. G. Rodewald, *J. Am. Chem. Soc.* **89**, 862 (1967).
27. R. Corriu, J. Massé, and C. Guérin, *J. Chem. Res. (S)* 160 (1977).
28. R. Corriu and B. Henner, *J. Organomet. Chem.* **102**, 407 (1975).
29. K. Fukui, *Bull. Chem. Soc. Jap.* **39**, 498 (1966).
30. P. Caramella and K. N. Houk, *J. Am. Chem. Soc.* **98**, 6397 (1976)
31. R. G. Pearson, *J. Am. Chem. Soc.* **85**, 3533 (1963); R. G. Pearson and J. Songstad, ibid. **89**, 1827 (1967).
32. G. Klopman in *Chemical Reactivity and Reaction Paths,* edited by G. Klopman (Wiley, New York, 1974), Chapter 4.
33. J. O'M. Bockris and A. K. N. Reddy, *Modern Electrochemistry* (Plenum, New York, 1973), Chapter 3.
34. C. Minot and N. T. Anh, *Tetrahedron Lett.* 3905 (1975); N. T. Anh, *Topics Curr. Chem.* **88**, 146 (1980); A. J. Parker, *J. Chem. Soc.* 1328 (1961).
35. A. Loupy and J. Seyden-Penne, *Tetrahedron Lett.* 2571 (1978).
36. J. Fossey and D. Lefort, *Tetrahedron* **36**, 1023 (1980).
37. N. D. Epiotis, *J. Am. Chem. Soc.* **94**, 1924 (1972).
38. J. A. Berson and L. Salem, *J. Am. Chem. Soc.* **94**, 8917 (1972).
39. J. A. Berson, *Acc. Chem. Res.* **5**, 406 (1972).
40. L. Salem in *Chemical and Biochemical Reactivity,* edited by E. B. Bergmann and B. Pullman (Israel Academy of Sciences and Humanities, Jerusalem, 1974), p. 329.
41. R. B. Woodward and R. Hoffmann, *The Conservation of Orbital Symmetry* (Verlag Chemie, Weinheim, 1970), pp. 114–118.
42. J. A. Berson and G. L. Nelson, *J. Am. Chem. Soc.* **89**, 5503 (1967); J. A. Berson, *Acc. Chem. Res.* **1**, 152 (1968).
43. R. F. W. Bader, *Molec. Phys.* **3**, 137 (1960); *Can. J. Chem.* **40**, 1164 (1962).
44. L. Salem, *Chem. Phys. Lett.* **3**, 99 (1969).
45. R. G. Pearson, *J. Am. Chem. Soc.* **91**, 1252 (1969); *Acc. Chem. Res.* **4**, 152 (1971).
46. L. Bartell, *J. Chem. Ed.* **45**, 754 (1968).
47. J. Burdett, *J. Chem. Soc. A*, 1195 (1971).
48. H. Metiu, J. Ross, R. Silbey, and T. F. George, *J. Chem. Phys.* **61**, 3200 (1974); see also Ref. 6 for a review.
49. U. Opik and M. H. L. Pryce, *Proc. Roy. Soc.* **A238**, 425 (1957).
50. D. H. W. den Boer, P. C. den Boer, and H. C. Longuet-Higgins, *Molec. Phys.* **5**, 387 (1962).
51. I. B. Bersuker, *Coord. Chem. Rev.* **14**, 357 (1975).
52. H. C. Longuet-Higgins, *Proc. Roy. Soc.* **A235**, 537 (1956).

53. R. G. Pearson, *Symmetry Rules for Chemical Reactions* (Wiley, New York, 1976), pp. 190 sqq.
54. A. Rauk, L. C. Allen and K. Mislow, *J. Am. Chem. Soc.* **94,** 3035 (1972).
55. A. Strich and A. Veillard, *J. Am. Chem. Soc.* **95,** 5574 (1973).
56. R. S. Berry, *J. Chem. Phys.* **32,** 933 (1960).
57. K. Nakamoto, *Infrared Spectra of Inorganic and Coordination Compounds* (Wiley, New York, 1963), Fig. II-23.
58. J. Wright and L. Salem, *J. Am. Chem. Soc.* **91,** 5947 (1969).
59. I. B. Bersuker, *Nouv. J. Chim.* **4,** 139 (1980).
60. H. B. Bürgi, *Inorg. Chem.* **12,** 2321 (1973).
61. H. B. Bürgi, J. D. Dunitz, and E. Shefter, *J. Am. Chem. Soc.* **95,** 5065 (1973).
62. K. Müller, *Angew. Chem. Internatl. Ed.* **19,** 1 (1980).
63. H. B. Bürgi, J. M. Lehn, and G. Wipff, *J. Am. Chem. Soc.* **96,** 1956 (1974); N. T. Anh and O. Eisenstein, *Nouv. J. Chim.* **1,** 61 (1977).
64. A. Streitwieser, *Molecular Orbital Theory for Organic Chemists* (Wiley, New York, 1961).
65. R. Bonaccorsi, E. Scrocco, and J. Tomasi, *J. Chem. Phys.* **52,** 5270 (1970).
66. R. Bonaccorsi, E. Scrocco, and J. Tomasi, *J. Am. Chem. Soc.* **98,** 4049 (1976).
67. J. Tomasi in *Quantum Theory of Chemical Reactions,* Vol. I, edited by R. Daudel, A. Pullman, L. Salem, and A. Veillard (Reidel, Dordrecht, 1980), p. 191.
68. M. Saunders, P. Vogel, E. L. Hagen, and J. Rosenfeld, *Acc. Chem. Res.* **6,** 53 (1973).
69. P. P. Dymerski, R. M. Prinsteen, P. F. Bente, and F. W. McLafferty, *J. Am. Chem. Soc.* **98,** 6834 (1976).
70. O. Chalvet, R. Constanciel, and J. C. Rayez in *Chemical and Biochemical Reactivity,* edited by E. D. Bergmann and B. Pullman (Israel Academy Sciences and Humanities, Jerusalem, 1974), p. 77.
71. G. Jackson and G. Porter, *Proc. Roy. Soc.* **A260,** 12 (1961).

7

The Role of Spin;
The Various Manifestations
of Electron Exchange

The energies of individual electrons, such as those that occupy frontier orbitals or the labile electrons involved in orbital correlation diagrams, do not always suffice to delineate the preferred path of a chemical reaction. The role of state correlations, in which repulsion between electrons may determine important modifications to the one–electron picture, has already been examined. The spins of electron pairs involved in bond formation and their relative "parallel" or "antiparallel" spin components are also a discriminating factor in chemical reactions. Triplet and singlet states that originate even from the same orbital configuration may have entirely different behavior (see the $^1\pi,\pi^*$ and $^3\pi,\pi^*$ states of ethylene in Fig. 3-5, or the unique behavior of the $^3\pi,\pi^*$ state in the cleavage of alkanones in Fig. 5-11). This difference can be ascribed to the so-called "exchange energy," a notion that we define in Section 7-1.

When speaking of "parallel" and "antiparallel" spins, it is useful to remember the vectorial schemes (Fig. 7-1) that this terminology covers.[1] For two electrons with a total spin angular momentum quantum number $S_{total} = 0$ the individual spin angular momentum vectors are indeed antiparallel. However, within the three cases where the combined quantum number $S_{total} = 1$, none corresponds to truly parallel spin angular momentum vectors.

7.1 Exchange: Molecular Orbital and Valence-Bond Concepts
Depending on the method being used, the term "exchange integral" covers two entirely different notions.

Molecular Orbital Definition
Consider first two *atomic* orbitals ϕ_A and ϕ_B. Their "overlap density" is defined by the product

$$\text{Overlap density} \equiv \phi_A\phi_B \tag{7-1}$$

and this is true regardless of whether the overlap integral

$$S_{AB} = \int \phi_A\phi_B \, d\tau = 0 \quad \text{or} \neq 0 \tag{7-2}$$

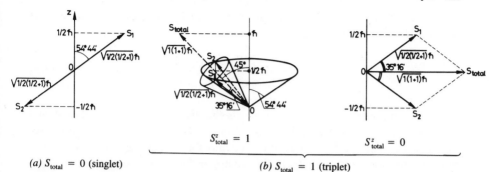

$S_{total}^z = 1$

$S_{total}^z = 0$

(a) $S_{total} = 0$ (singlet)

(b) $S_{total} = 1$ (triplet)

FIGURE 7-1. Vectorial schemes for "antiparallel" and "parallel" spins.[1] The spin vectors for the individual electrons are S_1 and S_2, and S_{total} refers to the combined spins. For $S_{total}^z = -1$, the scheme is the reverse of that for $S_{total}^z = 1$. Reprinted in part with permission from *Quantum Chemistry, An Introduction* by W. Kauzmann. Copyright 1957 by Academic Press.

By definition, the exchange integral "between" ϕ_A and ϕ_B or "for" ϕ_A and ϕ_B is the Coulombic interaction between the overlap density and itself:

$$K_{AB} = \int (\phi_A \phi_B)(1)\frac{1}{r_{12}}(\phi_A \phi_B)(2)d\tau_1 \, d\tau_2 \qquad (7-3)$$

This integral K_{AB} is an electrostatic self-energy of repulsion and is therefore always positive.

The size of K_{AB} depends to some extent on the value of S_{AB}, but more so on the compactness versus diffuseness of the overlap density. For two orthogonal orbitals on the same center ($S_{AB} = 0$), the value of K_{AB} depends on the type of orthogonality (radial or angular). Typical values are 0.99 eV for $1s$ and $2s$ orbitals of helium with effective nuclear charge 1.675, and 0.77 eV for $1s$ and $2p$ orbitals. For two atomic orbitals centered on adjacent atoms, the exchange integral falls off rapidly, approximately as the square of the overlap.

For two *molecular* orbitals, the definition of K_{ij} relative to ψ_i and ψ_j and their overlap density $\psi_i\psi_j$ is identical with the atomic orbital case. As given already in Chapter 1 (footnote *a* in Table 1-2):

$$K_{ij} = \int (\psi_i\psi_j)(1)\frac{1}{r_{12}}(\psi_i\psi_j)(2)d\tau_1 \, d\tau_2 \qquad (7-4)$$

Again, such an exchange integral between molecular orbitals is always positive.

Valence-Bond Definition

In the valence-bond picture, however, the exchange integral has an entirely different interpretation; its *sign* is also different. In the calculation of the energy of interaction between two hydrogen atoms (Section 1.2), the total Hamiltonian operator is generally divided into two parts, an unperturbed Hamiltonian and a perturbation Hamiltonian due to the approach of the atoms [see also (6-2a)]. This perturbation Hamiltonian, for an atom A (electron 1) that encounters an atom B (electron 2), is given by

$$H' = \frac{1}{R} - \frac{1}{r_{A2}} - \frac{1}{r_{B1}} + \frac{1}{r_{12}} \tag{7-5}$$

where R is the internuclear distance, r_{A2} is the distance between nucleus A and electron 2, and r_{12} is the interelectronic distance. The total energy expression for both ground singlet and excited triplet contains the expectation value of this operator for the wave function (1-2). This energy includes the cross-term

$$J = \int \phi_A(1)\phi_B(1) \, H' \, \phi_A(2)\phi_B(2) d\tau_1 \, d\tau_2 \tag{7-6}$$

which is called the *exchange integral*. In this integral the first and fourth terms of H' contribute positively, the latter through an exchange integral identical with the atomic orbital exchange integral K_{AB} (7-3). The nuclear–electron attraction parts of H', on the other hand, contribute negatively. Overall, J, which is the average of the perturbation Hamiltonian over the overlap density $\phi_A\phi_B$, is a *negative* quantity.

The valence-bond energy difference between excited triplet state and ground singlet state is given by[2,*]

$$E(^3H_2) - (^1H_2) = \frac{Q - J}{1 - S^2_{AB}} - \frac{Q + J}{1 + S^2_{AB}} \tag{7-7}$$

where Q is the "Coulomb" integral corresponding to the diagonal value of H' over $\phi_A(1) \, \phi_B(2)$ or $\phi_B(1) \, \phi_A(2)$:

$$Q = \int \phi_A(1) \, \phi_B(2) \, H' \, \phi_A(1) \, \phi_B(2) \, d\tau_1 d\tau_2 \tag{7-8}$$

Hence, to a first approximation, the valence-bond exchange integral measures, in its absolute value, the triplet-singlet energy difference:

$$-J \approx \tfrac{1}{2} (E_{triplet} - E_{singlet}) \tag{7-9}$$

This expression is used in Section 7.5 to incorporate J in a "phenomenological" Hamiltonian.

7.2   Hund's Rule and Why Triplets Are Lower than Singlets for the Same Orbital Configuration

Established in 1924 from the empirical observation of line spectra of atoms, Hund's rule states[3] that the spectroscopic "term" with the largest spin multiplicity lies

---

*The two terms on the RHS of (7-7) correspond respectively to the energies of wave functions (1-4) and (1-2).

lowest in energy and also that, for a given multiplicity, terms with the largest L (orbital angular momentum) values lie lowest in energy. Hund's rule applies to a given electronic configuration of an atom and a common example is the oxygen atom (configuration p^4 and states $^3P < {}^1D < {}^1S$). At first sight the rule also seems valid for molecules, with the O_2 molecule (configuration π^{*2} and states $^3\Sigma_g^- < {}^1\Delta_g < {}^1\Sigma_g^+$) as an illustration.

The traditional interpretation of Hund's rule, which discriminates in the simplest case between the singlet and triplet associated with a given open-shell electronic configuration $\psi_1\psi_2$, calls on electron exchange. The relative two-electron energies of the singlet and triplet are

$$\text{singlet:} \quad J_{12} + K_{12} \tag{7-10}$$
$$\text{triplet:} \quad J_{12} - K_{12}$$

where J_{12} and K_{12} are as defined in footnote a in Table 1-2. Let us consider the region defined by the overlap density $\psi_1\psi_2$ between the two orbitals (Fig. 7-2). The two electrons, which in the singlet state are free to occupy simultaneously this "exchange region," are forbidden to do so in the triplet state by the Pauli exclusion principle (see Section 4.1). A term corresponding to the repulsion between the electrons in this region thus occurs in the singlet state, whereas the total absence of such repulsion—a Coulombic "relief," in a sense—gives a term of opposite sign in the triplet.

Colpa and his collaborators[4-61] were the first to show that the previous explanation is too simple-minded.* They pointed out that (7-10) has profound consequences on the orbitals themselves. Both orbitals tend to distort so as to decrease the size of the two-electron energy J_{12}, which is large and positive in either state. In the singlet an overall expansion decreases both J_{12} and the unfavorable K_{12} term. In the triplet the tendency for orbital expansion to decrease J_{12} is counteracted by a tendency, albeit weaker, for orbital contraction to increase the favorable K_{12} term. Overall, the orbital relaxation has a much more drastic effect on the singlet than on the triplet, as shown in scheme (7-11). Although both states

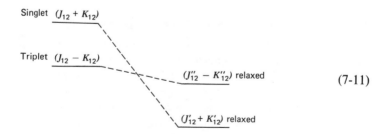

Singlet $(J_{12} + K_{12})$

Triplet $(J_{12} - K_{12})$

$(J_{12}'' - K_{12}'')$ relaxed

$(J_{12}' + K_{12}')$ relaxed

$$\tag{7-11}$$

are stabilized, the singlet state now lies below triplet!

Why, then, is Hund's rule operative? The final answer lies in the *electron-nuclear attraction* of the relaxed states. The triplet, with more contracted orbitals than the singlet, has a larger nuclear-electron attraction, and finally a more negative

*For a review, see Ref. 7.

Singlet   Possible   Possible   Possible
Triplet   Impossible   Impossible   Impossible

FIGURE 7-2.   Overlap density (dotted, positive; crosses, negative) for *sp* configuration and comparison between electron pair situations in singlet and triplet states. Orbital amplitudes are either positive (white) or negative (shaded). Electrons are shown as dots.

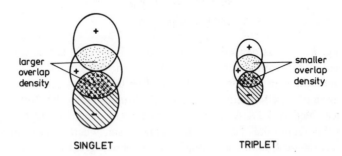

FIGURE 7-3.   Differential orbital relaxation in singlet and triplet states for *sp* configurations.

total electronic energy. The simple picture of Hund's rule as originating directly from electron exchange is lost. The correct picture is one in which electron correlation distorts singlet and triplet orbitals in a differential manner (Fig. 7-3). This differential deformation—which, true, does find its origin in exchange terms of opposite sign—produces in turn a larger nuclear attraction in the triplet, the more stable state.

7.3   Violation of Hund's Rule and Dynamic Spin Polarization

The ''static'' polarization of an electron spin by a distinct unpaired spin in a molecular radical has been known and understood for about 25 years.[8-11] It occurs in the CH bonds of aromatic radicals. Because of Hund's rule, which applies locally to the orthogonal π and σ orbitals at the carbon atom center, the σ electron closest to an unpaired π electron tends to have its spin *parallel* to that of the π electron. Likewise, static polarization arises in the π system of conjugated radicals. For instance, in the presence of the odd electron spin at the carbon termini of the allyl skeleton, the two bonding electrons take on different distributions: the partner with the same spin as the unpaired electron prefers to be at the ends of the chain (Fig. 7-4). These effects are at the origin of electron–spin–resonance proton hyperfine-spitting interactions in aromatic hydrocarbon radicals.

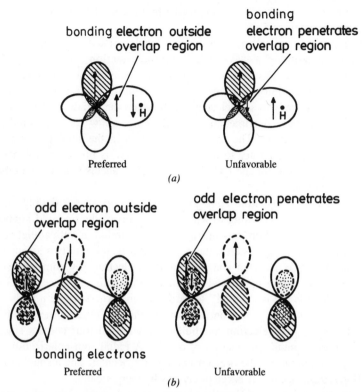

FIGURE 7-4. Static spin polarization in molecules: *(a)* CH bond in π radical; *(b)* allyl radical. In the preferred situation, the odd electron and a bonding electron are in a local triplet configuration. In the unfavorable cases the electrons (in a local singlet configuration) are drawn with either the odd electron or a bonding electron occupying the overlap density region (they may of course occupy this region together). In the allyl radical, bonding molecular orbitals are drawn in dotted lines and nonbonding orbitals, in full lines. Overlap density regions are dotted (positive) or with crosses (negative).

In Section 7.2 we showed, however, that extremely careful language is required to describe static spin polarization: the odd-electron spin does not really "pull" an electron of parallel spin toward it. *The excess of parallel spin simply reflects the energetically unfavorable situation if an electron of opposite spin momentum were to come nearby.* The electrons would then pair up in the exchange region, and large Coulomb repulsions would follow.

Let us now consider some remarkable cases where Hund's rule is violated in molecules. A few such violations have been known in atoms (^1D below ^3D for the $3s3d$ configuration of magnesium) and explained by a degeneracy of electronic configurations ($3s3d$ and $3p^2$ in the singlet manifold).[12] But two molecular examples have come to the forefront in recent years: that of 90°–twisted ethylene (Section 3.4)[13] and that of square-planar cyclobutadiene (Section 3.9).[14] In the first case the singlet is more stable than the triplet by roughly 0.1 eV and in the second case by approximately 0.5 eV.

Kollmar and Staemmler[15] have provided a lucid explanation of this violation of Hund's rule, in terms of a "dynamic" spin polarization effect. This effect is illustrated in Fig. 7-5. Essentially, the molecule seeks out a local instantaneous (π,σ) triplet configuration at each carbon atom as in the static polarization effect. This automatically leads to singlet coupling between the diradical electrons in the orthogonal $p\pi$ orbitals. A triplet diradical does not allow for two simultaneous local triplet configurations. The overall polarization in the singlet state is called *dynamic*. Indeed, and contrary to the static case there is no *net* spin density; the local triplet polarization for the CC bond ↑↓ singlet cancels that for the other CC bond singlet component ↓↑.

In cyclobutadiene, dynamic spin polarization occurs in the doubly occupied a_{2u} orbital* and varies according to whether the spins of the two odd nonbonding electrons (in the *lozenged* orbitals in Fig. 3-7) are parallel or antiparallel. If these spins are antiparallel (singlet state), the a_{2u} orbital can split into two lozenged-type orbitals, one for its α electron and one for its β electron, by admixing some π character from the empty b_{1u} orbital (Fig. 7-6). Physically, it is easy to conceive that one a_{2u} electron (α) is "pulled" toward the two corners where it finds its own odd spin and the other a_{2u} electron (β), toward the two other corners. But in the triplet state one a_{2u} (α) electron is "pulled" toward all four corners since both lozenged orbitals contain an electron of α spin. A radial spin polarization, which gives a large amplitude for the α electron on all four centers and a small amplitude for the β electron, occurs. It is relatively costly energetically.

In summary, violations of Hund's rule arise from dynamic spin polarization, an instantaneous electron correlation effect in which electrons of inner-shell orbitals correlate their spins with those of the odd electrons. It is an *interorbital* correlation[17] effect that decreases the overall Coulomb repulsion energy of singlet states.

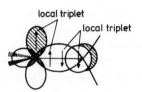

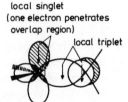

FIGURE 7-5.   Dynamical polarization in 90°–twisted ethylene: instantaneous correlations of the electron spins. Overlap density regions are more darkly shaded. *(left)* Singlet diradical state. *(right)* Triplet diradical state.

*See Refs. 14 and 16 (Fig. 2*b* of the latter article seems to have the radial polarizations of the α and β electrons inverted).

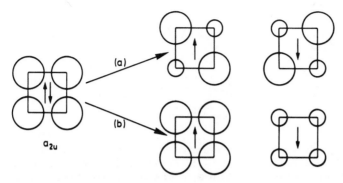

FIGURE 7-6.   Dynamic spin polarization of the inner bonding π shell of cyclobutadiene. The *outer*, odd, non-bonding electrons are *(a)* singlet-paired and *(b)* triplet-paired with two ''up'' spins.[16]

7.4 Spin Density Waves of Various Types

In Section 7.3 we saw that Hund's rule—which arises from differential electron correlation between singlet and triplet with identical orbital configurations—gives rise to a specific alignment of spins in molecules, even in their singlet states (Fig. 7-5). The overall net spin density vanishes, and so does the net spin density at any point in space. But the fine analysis shows that this vanishing density is due to the symmetry-required combination of two equivalent situations that each, separately, have strong spin polarization. For instance, the singlet state of 90°-twisted ethylene can be described as

$$(7\text{-}12)$$

in which the CH_2 regions are all strongly spin polarized. In either of the two contributors to the state the ''up'' spins and ''down'' spins occupy different parts of space—*spin waves*, as it were (Fig. 7-7a), with a net spin density at each point in space.

Such spin waves were first introduced by Löwdin[18] to describe conjugated molecules so as to allow for correlation between antiparallel spins. Orbitals for the $\alpha(\beta)$ spins are constructed so as to be largely concentrated on even-numbered atoms; those for the $\beta(\alpha)$ spins are localized mainly on the odd-numbered atoms. To these ''alternant'' orbitals correspond alternant spin waves,[19] as shown in Fig. 7-7b. The lowest Hartree-Fock state, in which the electrons are paired in orbitals with amplitude monotonically varying along the carbon chain, is unstable relative to this new state in which the α and β spins occupy respectively the two different alternant orbitals (for such ''triplet instability,'' see Section 3.1). The correct symmetry-adapted ground state is a linear combination of the two possibilities, that with α spins on even atoms and that with α spin on the odd atoms.

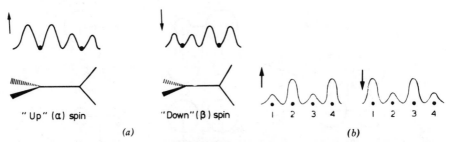

FIGURE 7-7.   Spin waves: *(a)* in 90°-twisted ethylene [amplitude along the H_2CCH_2 skeleton is shown schematically in one dimension, see eq. (7-12)]; *(b)* in linear chain ("alternant" spin waves).[19] Part *(b)* reprinted with permission from D. B. Abraham and A. D. McLachlan, *Molec. Phys.* **12**, 319 (1967). Copyright 1967 by Taylor and Francis.

Yamaguchi[20,21] has made an extensive study of spin waves in organic radicals by using the Heisenberg Hamiltonian, which we study in Section 7.5, as well as molecular orbital theory.[22] For cyclic π-radicals such as the cyclopentadienyl radical, the ground spin structure is shown to be "torsional" or "helical"[23–25] (Fig. 7-8). Relative to the *z* axis that carries the spin on the first atom, the spin on the second atom is directed at an angle of $2\pi/5$, that on atom 3 at $4\pi/5$ (or $-\pi/5$), that on atom 4 at $6\pi/5$ (or $\pi/5$), and that on atom 5 at $8\pi/5$ (or $-2\pi/5$). The sum of the vectors, which are all in the same plane, is zero. A similar situation rises in the tetrahedral H_4 diradical, with again the four spin vectors adding up to zero (Fig. 7-9).

These special torsional spin density waves correspond to an "instability" of the normal restricted Hartree-Fock wave function (1-23) for the ground doublet state when monoexcited configurations are mixed into it.[22,†] An unrestricted Hartree-Fock solution of type (1-22) is preferred. Physically, the odd spin in the nonbonding orbital polarizes the electron pairs in the inner shells. If the configurations that render (1-23) unstable involve excitation (electron from occupied to empty orbital) *without* spin flip the orbitals in (1-22) are "different orbitals for different spins" similar to alternant orbitals. The α electrons and their β counterparts (formerly

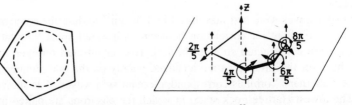

FIGURE 7-8.   Cyclopentadienyl radical and its torsional spin waves.[20] (The spin vectors on the RHS all lie in the same plane and add to zero.)

†Brillouin's theorem[26] requires that the Hartree-Fock solution be stable to a small admixture of monoexcited configurations.

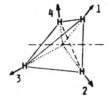

FIGURE 7-9. Spin vectors in the tetrahedral H_4 system.[20] Reprinted with permission from K. Yamaguchi, *Chem. Phys. Lett.* **30**, 288 (1975). Copyright 1975 North-Holland Publishing Company.

paired) now occupy partially different segments of space. For a three-electron three atom chain (Fig. 7-10*a*), these "M_z-modulated" spin orbitals are given by

$$\chi_1 = [\cos(\tfrac{1}{2}\lambda)\psi_1 + \sin(\tfrac{1}{2}\lambda)\psi_3]\alpha$$

$$\chi_1' = [\cos(\tfrac{1}{2}\lambda)\psi_1 - \sin(\tfrac{1}{2}\lambda)\psi_3]\beta \qquad (7\text{-}13)$$

$$\chi_2 = \psi_2$$

Here ψ_1, ψ_2, and ψ_3 are the ordinary allylic molecular orbitals for the chain. Orbitals χ_1, χ_1', and χ_2 are analogous to alternant spin density waves.

If, on the other hand, the restricted Hartree-Fock function is unstable to excitations *with* spin flip (electron, with same spin as the odd electron, from occupied to empty orbital together with spin reversal), then the corresponding spin-orbital solutions in (1-22) are *mixtures* of α-spin orbital fragments and β-spin orbital fragments.[22] For the three-atom chain, we obtain

$$\mu_1 = [\cos(\tfrac{1}{2}\lambda)\psi_1]\alpha + [\sin(\tfrac{1}{2}\lambda)\psi_3]\beta$$

$$\mu_1' = [\cos(\tfrac{1}{2}\lambda)\psi_1]\beta + [\sin(\tfrac{1}{2}\lambda)\psi_3]\alpha \qquad (7\text{-}14)$$

$$\mu_2 = \psi_2\frac{(\alpha + \beta)}{\sqrt{2}}$$

The corresponding spin density waves are called "M_x-modulated"[22]; the wave

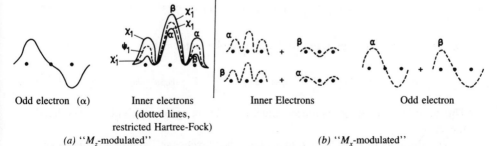

| Odd electron (α) | Inner electrons (dotted lines, restricted Hartree-Fock) | Inner Electrons | Odd electron |

(a) "M_z-modulated"   *(b)* "M_x-modulated"

FIGURE 7-10. Different types of spin density wave for a three-electron three atom chain.

function must be projected on to S_z to recover the proper commutation properties of Hamiltonian and spin operators.

The most general orbitals are *combinations*[27] of M_x- and M_z-modulated spin density waves, which not surprisingly are "helical" or "torsional" spin density waves similar to those shown both in Figs. 7-8 and 7-9. The spin density modulates in both x and z directions. It must be remembered, however, that such solutions are not proper eigenfunctions of S^2. This explains, for instance, why a full spin does not appear to be recovered in the cyclopentadienyl radical; therefore, one must be extremely wary before giving physical reality to these torsional solutions.

7.5   Heisenberg Hamiltonian and Applications

Although the interactions between two odd electrons are purely electrostatic (there are tiny magnetic effects), the result (7-9) from Section 7.1, in which the energy difference between lowest triplet and singlet states depends only on the spin multiplicity, has encouraged authors to write out "formal" or "phenomenological" Hamiltonians depending on the spin alone. The most famous such phenomenological Hamiltonian is due to Heisenberg,[28] Dirac,[29] and Van Vleck.[30] For a set of radicals each initially in a *doublet* state the interaction Hamiltonian is written out as

$$\mathcal{H} = -2 \sum_{i > j} J_{ij} \, \mathbf{S}_i \cdot \mathbf{S}_j \qquad (7\text{-}15)$$

where J_{ij} is the negative valence-bond exchange integral between the odd electrons i and j on different radical centers (orbitals unspecified), with respective spin angular momentum operators $\mathbf{S}_i$ and $\mathbf{S}_j$.

For two electrons we express the operator $\mathbf{S}_1 \cdot \mathbf{S}_2$ as

$$-2J\mathbf{S}_1 \cdot \mathbf{S}_2 = -J(\mathbf{S}^2 - \mathbf{S}_1^2 - \mathbf{S}_2^2) \qquad (7\text{-}16)$$

where $\mathbf{S} = \mathbf{S}_1 + \mathbf{S}_2$ is the operator for the total spin angular momentum, with eigenvalue S. For two spins $\mathbf{S}_1$ and $\mathbf{S}_2$ with eigenvalues $S_1 = \frac{1}{2}$, $S_2 = \frac{1}{2}$, the expectation value of $\mathcal{H}$ is

$$E = -J[S(S + 1) - S_1(S_1 + 1) - S_2(S_2 + 1)] \qquad (7\text{-}17)$$

The energy difference between triplet and singlet states is thus

$$E_{\text{triplet}} - E_{\text{singlet}} = -J[S_{\text{triplet}}(S_{\text{triplet}} + 1) - S_{\text{singlet}}(S_{\text{singlet}} + 1)] \qquad (7\text{-}18)$$

This difference in energy is indeed equal to $-2J$, in agreement with valence-bond theory.

The Heisenberg Hamiltonian gives only the "exchange" splitting between singlet and triplet, not their correct total energies, which are calculated to be $(3/2)J$

and $-(1/2)J$, respectively (compare with 7-7). It is possible to write out a full spin Hamiltonian that does yield these energies by simply adding to (7-15) an appropriate constant term. For two electrons, this corrected spin Hamiltonian is[31]

$$\mathcal{H}_{spin} = Q - J (\tfrac{1}{2} + \mathbf{S}_1 \cdot \mathbf{S}_2) \tag{7-19}$$

which gives an energy $Q + J$ for a spin singlet function and an energy $Q - J$ for a spin triplet eigenfunction. At all times· it should be remembered that although Hamiltonians (7-15) and (7-19) suggest a true vector coupling of electron spins, there is no such physical coupling in reality. The Heisenberg Hamiltonian can be improved in the case of nonorthogonal orbitals on two radical centers[31,32] or in the case of radicals that carry orbital angular momentum in addition to their spin angular momentum.[33]

Yamaguchi[20] has given an illuminating application of the Heisenberg Hamiltonian to the determination of allowed spin structures for three odd electrons. He considered a triangular system of three atoms each carrying an odd spin vector—as in a radical addition reaction. The model and coordinate axes for the spin vectors are shown in Fig. 7-11. Each spin $\mathbf{S}_i$ is assumed to be a *classical* vector* carried by a unit vector $\boldsymbol{\mu}_i$ and with magnitude S_i. The axis carrying $\mathbf{S}_3$ for the third electron is chosen as z axis, and the other two spin vectors are placed in the xz plane. Assuming the energy of the system to arise purely from exchange—in the valence-bond sense—we have, from (7-15),

$$E = -2S^2[J \cos (\theta_2 - \theta_1) + J_1 \cos \theta_1 + J_1 \cos \theta_2] \tag{7-20}$$

where S is the common magnitude of all three spins (note that S is the magnitude of a localized spin and is not related to the spin operator $\mathbf{S}^2$). Following Yamaguchi, we state that the energy be stationary with respect to small variations of θ_1 and θ_2, whence

$$J \sin (\theta_2 - \theta_1) - J_1 \sin \theta_1 = 0 \tag{7-21}$$
$$-J \sin (\theta_2 - \theta_1) - J_1 \sin \theta_2 = 0$$

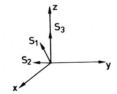

FIGURE 7-11.   Triangular three-atom system (exchange integrals) and coordinate axis for the spin-vectors. All three spin-vectors are chosen to lie in the xz plane.[20]

---

*In this classical Heisenberg model, the spin basis functions $\alpha_1\beta_2\alpha_3$, $\beta_1\alpha_2\alpha_3$ and $\alpha_1\alpha_2\beta_3$ are not eigen functions of the operator $\mathbf{S}^2$. In the earlier, quantal Heisenberg model of Ref. 21, the spin functions are properly combined to be eigen functions of $\mathbf{S}^2$ and $\mathbf{S}_z$.

Equations (7-21) can be solved by successive addition and subtraction. The first solution is

$$\sin \theta_1 = -\sin \theta_2 = 0, \quad \text{giving} \begin{cases} \theta_1 = \pi, \theta_2 = 0(I) \text{ and } \theta = 0, \theta = \pi(II) \\ \theta_1 = \theta_2 = \pi(III) \text{ or } 0(IV) \end{cases}$$

$$(7\text{-}22a)$$

The corresponding spin arrangements are shown in Fig. 7-12 (I–IV). All the spin vectors are carried by the z axis; the spin structures are "axial" M_z-modulated spin density waves as discussed in Section 7.4. In I and II atom 3 interacts with the singlet molecule 12 to give a doublet state; in III and IV it interacts with the triplet molecule to give a second doublet state and a quartet. The second solution is

$$\sin \theta_1 = -\sin \theta_2; \cos \theta_1 = \cos \theta_2 = -\frac{J_1}{2J} > \left(\left| \frac{J_1}{2J} \right| \leq 1 \right) \quad (7\text{-}22b)$$

For the equilateral triangle $(J_1 = J)$, this gives the degenerate solutions $\theta_1 = 2\pi/3$, $\theta_2 = 4\pi/3$, and $\theta_1 = 4\pi/3$, $\theta_2 = 2\pi/3$. The spin vectors no longer point along the z axis. The corresponding spin structures are "torsional" spin density waves of the same type as those in the cyclopentadienyl radical (Fig. 7-8): see V and VI in Fig. 7-12.

The general case of four spin vectors has also been treated;[34] applications of the method to the actual energetics of three-electron systems are given in Section 7.6.

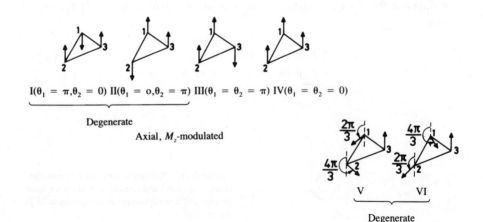

$I(\theta_1 = \pi, \theta_2 = 0)$ $II(\theta_1 = 0, \theta_2 = \pi)$ $III(\theta_1 = \theta_2 = \pi)$ $IV(\theta_1 = \theta_2 = 0)$

Degenerate

Axial, M_z-modulated

V   VI

Degenerate
Torsional (cf. Fig. 7-8)

FIGURE 7-12.   Spin arrangements for the three-electron triangular system.[20]

7.6 An Elementary Treatment of Free-Radical Reactions

In Section 7.5 we pointed out that the triangular approach of an odd-electron center to a diatomic molecule, each atom of which also carries an odd electron, is a simple model for the addition of a radical to a double bond

$$\begin{array}{c} \text{A} \\ \parallel \\ \text{B} \end{array} + \text{R·} \qquad (7\text{-}23)$$

Isosceles Triangular Approach

We can thus pursue the classical Heisenberg model one step further by calculating the actual energies corresponding to the spin structures shown in Fig. 7-12. In the particular case of an isosceles triangle with C_{2v} symmetry, it is legitimate to perform this calculation separately for the first doublet (I and II), the second doublet (III), and the quartet (IV) because there are no off-diagonal elements of the Hamiltonian between the two doublets. Let

$$\frac{J_1}{J} = x \qquad (7\text{-}24)$$

be the single parameter of this isosceles triangular approach in which the reactant does not give any preference to either side of the molecule. From (7-20), (7-22), and (7-24) the energy solutions are[20] (with the contributions from the different resonance structures):

$$-\frac{E_1}{2JS^2} = -1 \qquad \text{(structures I} \leftrightarrow \text{II)}$$

$$-\frac{E_2}{2JS^2} = -2x + 1 \quad \text{(structure III)}$$

$$-\frac{E_3}{2JS^2} = +2x + 1 \quad \text{(structure IV; quartet)} \qquad (7\text{-}25)$$

$$-\frac{E_4}{2JS^2} = -\frac{1}{2}x^2 - 1 \quad \text{(structures V} \leftrightarrow \text{VI)}$$

The corresponding energy plots are shown in Fig. 7-13 (only $x > 0$ values are

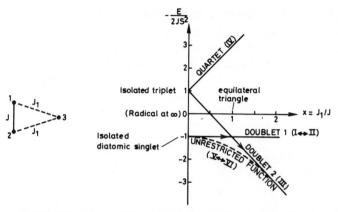

FIGURE 7-13.   Functional dependence of the classical Heisenberg energies for *isosceles* triangular approach of center 3 to bond 1—2.[20,27] For structures, see Fig. 7-12. Compare with Fig. 7-14 (*colinear, quantum Heisenberg model*). Reprinted with permission from K. Yamaguchi, *Chem. Phys. Lett.* **30,** 288 (1975). Copyright 1975 by North-Holland Publishing Company.

considered, since the valence-bond exchange integral is negative except in very rare cases, such as two orthogonal atomic orbitals on the same center).*

For $x = 0$, we obtain the energies of the isolated molecule in its singlet and triplet states, respectively. As the odd radical center approaches, there is little interaction with the singlet (one unfavorable pairing, and one favorable pairing). The ground doublet state corresponds initially to state 1, formed by the encounter of the singlet with the odd atom. But the triplet, with its two spins directed in a favorable pairing manner relative to the odd electron, is strongly stabilized. Hence the higher doublet state 2, formed by the encounter of the triplet with the odd atom, is strongly stabilized and soon becomes the ground state. The reaction can therefore proceed with apparent zero activation energy (along 1 and then along 2), with the odd electron coupling to the singlet molecule (I and II) at long distances, and then to the triplet (III) at shorter distances. Figure 7-13 shows that an even better solution for the system is to adopt, at the outset, the torsional spin structure V or VI, which then remains the ground state throughout the reaction (dashed line, Fig. 7-13). This means that an unrestricted Hartree-Fock function, built on orbitals similar to (7-14)—but adapted to the triangular chain[27]—gives a slightly better energy than those (I–III) obtained from restricted Hartree-Fock functions. This UHF function, however, has the limitations mentioned at the end of Section 7.4.

A full valence-bond approach[35] to this same triangular radical addition reveals a significant barrier along the first portion of the pathway until the surface crossing between the two doublets. This barrier does not appear in the simple Heisenberg treatment.

*Note that for $x = 0$, the molecular triplet III has an energy $-2JS^2$ and the molecular singlets I and II, $+2JS^2$, instead of $-(1/2)J$ and $(3/2)J$ as given by (7-17).

Colinear Approach

The Heisenberg model can also be applied* to the colinear approach of a radical center 3 to a molecule 12, as would occur in a radical abstraction reaction. However, even if the differentiation of (7-20) can be used to obtain the spin structures, the existence of off-diagonal matrix elements of $\mathcal{H}$ between the corresponding spin eigenfunctions does not allow (7-20) to be used for calculating the energies. These require diagonalization of the full Hamiltonian matrix between the different spin structures. For the model approach shown in Fig. 7-14, the results are[21]

$$-\frac{E_1}{J} = -\frac{1}{2}(1 + x) - (x^2 - x + 1)^{1/2}$$

$$-\frac{E_2}{J} = -\frac{1}{2}(1 + x) + (x^2 - x + 1)^{1/2}, \qquad x = \frac{J_1}{J} \qquad (7\text{-}26)$$

$$-\frac{E_3}{J} = \frac{1}{2}(1 + x)$$

The contribution of the different spin structures to the three states and the functional dependencies of the energies are shown in Fig. 7-14. The abstraction reaction should proceed freely in the ground doublet state. A more sophisticated calculation reveals[35] a slight activation barrier at $x = \frac{1}{2}$.

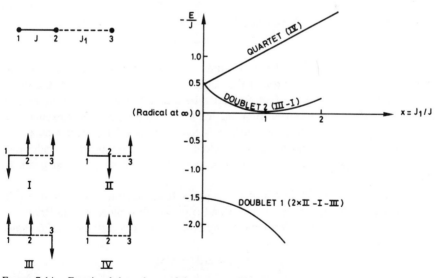

FIGURE 7-14. Functional dependence of the quantum Heisenberg energies for the *colinear* approach of center 3 to bond 1—2.[21] Also shown are the numerical contributions of the spin structures (LHS) to each state. Compare with Fig. 7-13 (*triangular*, classical Heisenberg model).

*See the quantum Heisenberg model in Ref. 21 and our footnote p. 199.

Finally, it should be noted that the various spin structures studied in Figs. 7-12 and 7-14 can be used to draw out state correlation diagrams[36] for free-radical reactions. The guiding principle is that the "magnetic point group" of the spin structures be conserved throughout the reaction path.

7.7   Superexchange

In addition to direct exchange effects between unpaired electrons, indirect exchange can arise through intervening electron pairs. Such "superexchange" has been known since 1934[37] to govern the interaction between two paramagnetic cations with intervening diamagnetic anions in magnetic insulators. Even if the direct interaction between the two paramagnetic centers is negligible, the presence of the intervening diamagnetic ion(s) can result in an appreciable exchange splitting between the two centers [38]; the spins on the two centers become coupled.

The simplest model for such superexchange involves the coupling of two odd electrons through an intervening electron pair ("three-center, four-electron model"). This model is illustrated in Fig. 7-15. Kramers' early description uses a Heisenberg-type Hamiltonian as a perturbation that couples the initial state i (terminal spins antiparallel) with the final state (terminal spins exchanged) through two intermediate excited states:

$$
\begin{array}{ccc}
\text{A} & \text{C} & \text{B} \\
\uparrow & \uparrow\downarrow & \downarrow \qquad \text{initial state } i \\
{}^{\ominus}\uparrow\downarrow & {}^{\oplus}\uparrow & \downarrow \qquad \text{excited state } \mu \\
{}^{\ominus}\uparrow\downarrow & {}^{\oplus}\downarrow & \uparrow \qquad \text{excited state } \mu' \\
\downarrow & \uparrow\downarrow & \uparrow \qquad \text{final state } f
\end{array}
\tag{7-27}
$$

In Kramers' theory superexchange requires the simultaneous intervention of both excited states μ and μ' (through the coupling $i \rightsquigarrow \mu \rightsquigarrow \mu' \rightsquigarrow f$; the direct couplings: $i \rightsquigarrow \mu'$ or $\mu \rightsquigarrow f$ are forbidden because the states differ by more than a direct transfer of an electron from one ion to a neighbor or a direct exchange of two electrons on a neighboring pair). It is thus a third-order perturbation effect. The inclusion of polar excited configurations also lies at the heart of Anderson's treatment of superexchange[39]; a full valence-bond treatment has been given by Jansen.[40]

The molecular orbital description of superexchange requires first a study of the direct exchange between two centers A and B.[32,41] Following Kahn and Briat[32] we start with the exact expression for the singlet-triplet energy difference as obtained from (7-7):

$$
\Delta E_{3-1} = -\frac{2(J - QS_{AB}^2)}{1 - S_{AB}^4}
\tag{7-28}
$$

FIGURE 7-15.   The three-center, four-electron model for superexchange.

where J and Q are defined in (7-6) and (7-8), respectively. Introducing the molecular orbitals

$$\psi_+ = \frac{\phi_A + \phi_B}{\sqrt{2(1 + S_{AB})}}$$

$$\psi_- = \frac{\phi_A - \phi_B}{\sqrt{2(1 - S_{AB})}}$$

$$(7-29)$$

Kahn and Briat show that the energy difference can be transformed into an expression that depends only on familiar molecular orbital quantities:

$$\Delta E_{3-1} = -\frac{2(J_{AB}^{MO} - K_{AB}^{MO} S_{AB}^2)}{1 - S_{AB}^4} + \frac{2S_{AB}(\epsilon_- - \epsilon_+)}{1 + S_{AB}^2} \qquad (7-30)$$

In (7-10) J_{AB}^{MO} is the two–electron molecular orbital type Coulomb integral over ϕ_A and ϕ_B and K_{AB}^{MO} the two-electron molecular orbital type exchange integral (see footnote a in Table 1-2); ϵ_- and ϵ_+ represent the usual Hartree-Fock type orbital energies, respectively, of ψ_- and ψ_+ [see (1-19)].

The molecular orbital expression for the energy difference has two opposing contributions: (1) a negative term (the Coulomb integral J_{AB}^{MO} is *positive*) that would lead to a triplet ground state (it arises from two-electron terms alone); and, (2) a positive term (for two centers A and B the antibonding energy ϵ_- always lies above the bonding energy ϵ_+) that leads to the normal singlet ground state. It is essentially proportional to the atomic orbital overlap. We now see how misleading are the valence-bond expression $\Delta E_{3-1} \approx -2J$ and the Heisenberg Hamiltonian $\mathcal{H} = -2J\mathbf{S}_1 \cdot \mathbf{S}_2$ since they both tend to convey a two-electron origin for the triplet-singlet energy difference. Equation (7-30) shows that the one-electron overlap is the main factor, as is well known from simple molecular orbital considerations.

Let us now turn to the four-electron three-center case.[42] Figure 7-16 shows a model linear system of two identical metal ions carrying an odd electron in a d_{z^2} orbital bridged by a ligand with a valence s orbital. Let Δ be the large positive energy difference between d_A, d_B, and s. Orbitals d_A and d_B, through their overlap with s, mix into themselves a small amount of s character, in a negative antibonding manner:

$$\left\{ \begin{array}{ll} d_A' = d_A + \dfrac{\beta}{\Delta} s, & \beta = \langle d_A |\mathcal{H}| s \rangle \\[3mm] d_B' = d_B + \dfrac{\beta}{\Delta} s, & = \langle d_B |\mathcal{H}| s \rangle < 0 \end{array} \right. \qquad (7-31)$$

where $\mathcal{H}$ is the total Hamiltonian. Charlot and Kahn now consider[42] these perturbed orbitals as "magnetic orbitals" that can be used to treat the two-center problem for ions A and B, as done previously. The overlap integral for A and B becomes

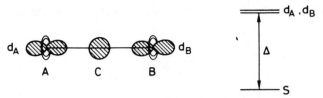

FIGURE 7-16.   Two model ions (d_{z^2} orbitals) bridged by a nonmetallic center.

$$S'_{AB} = \int d'_A d'_B d\tau \approx \frac{\beta}{\Delta} (S_{d_A,s} + S_{d_B,s}) \qquad (7\text{-}32)$$

This integral need no longer be positive if there are several intervening orbitals [see further in (7-31a)]. Similarly, in the presence of the bridging center, the energy difference $\epsilon'_- - \epsilon'_+$ now refers to the molecular orbitals $\psi'_- = d_A - d_B$ and $\psi'_+ = d_A + \dfrac{\beta}{\Delta} s + d_B$, which are fully delocalized over the intervening orbital. The second term of equation (7-30) is then used with these new quantities to determine the sign of ΔE_{3-1}.

7.8 Ferromagnetism and Antiferromagnetism in Binuclear Bridged Complexes

The molecular orbital theory of indirect exchange is a guide to the understanding of ferromagnetism versus antiferromagnetism in binuclear and polynuclear bridged complexes.ᐟ Typical such complexes involve metal ions doubly bridged by lone-pair carrying atoms such as oxygen and nitrogen. Examples are the nuclear complexes[43–45]

$$(7\text{-}33)$$

$$(7\text{-}34)$$

the heteronuclear complex[46]

$$(7\text{-}35)$$

and the polynuclear copper oxalate[47]

$$
\text{(structure)} \tag{7-36}
$$

Predictions of magnetic character rely on equation (7-30) for the triplet-singlet energy difference applied to the "magnetic orbitals" (7-31) of the metal ions, each partially delocalized toward the ligands surrounding the ion. The crucial quantities are the overlap S'_{AB} (7-32), the two-electron repulsion integral J'_{AB}

$$
J'_{AB} = \int d'^2_A (1) \frac{1}{r_{12}} d'^2_B (2) d\tau_1 \, d\tau_2 \tag{7-37}
$$

and the energy difference $(\epsilon'_- - \epsilon'_+)$ between the fully delocalized molecular orbitals built over all four interacting centers. In such metallic systems the (normal) ground singlet situation, in which the metallic spins alternate up and down, is said to be *antiferromagnetic*. A ground triplet, implying parallel spins throughout the chain, is said to be *ferromagnetic*. In practice, any ferromagnetic character is likely to come, not from a large two-electron integral J'_{AB}, but from cases where the one-electron "antiferromagnetic" term $2S'_{AB} (\epsilon'_- - \epsilon'_+)/(1 + S'^2_{AB})$ of (7-30) is very small (due to a small overlap or a small energy gap).

For compound (7-33), the half-empty Copper orbitals are $d_{x^2 - y^2}$ orbitals. They overlap with two different bridging ligand orbitals; the corresponding in-phase and out-of-phase molecular orbitals are shown in Fig. 7-17. For a copper-oxygen-copper angle of 104°, the compound is antiferromagnetic ($\Delta E_{3-1} = 509$ cm^{-1}), but for an angle of 96°, it is ferromagnetic ($\Delta E_{3-1} = -172$ cm^{-1})[43]. Indeed, as the angle closes, the overlap between $d_{x^2-y^2}$ and bridging p orbitals increases, stabi-

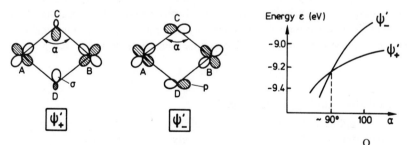

FIGURE 7-17. Symmetric and antisymmetric molecular orbitals for the four-center Cu⟨O⟩Cu system; energy behavior as a function of angle α.[44] Reprinted with permission from M. F. Charlot, S. Jeannin, Y. Jeannin, O. Kahn, J. Lucrèce-Abaul, and J. Martin-Frère, *Inorg. Chem.* **18**, 1675 (1979). Copyright 1979 by American Chemical Society.

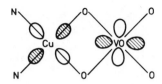

FIGURE 7-18.   Metallic orbitals in compound (7-35).

lizing ψ'_- relative to ψ'_+ (Fig. 7-17). The energy gap $\epsilon'_- - \epsilon'_+$ becomes very small, and the "ferromagnetic" two-electron term in (7-30) predominates[44]. At even smaller angles $\epsilon'_- - \epsilon'_+$ changes sign, but this is accompanied by a simultaneous change in sign of S'_{AB}, between the two "magnetic" orbitals

$$d'_A = d_A + \frac{\beta_\sigma}{\Delta_\sigma}\sigma - \frac{\beta_p}{\Delta_p}p, \qquad d'_B = d_B + \frac{\beta_\sigma}{\Delta_\sigma}\sigma - \frac{\beta_p}{\Delta_p}p \qquad (7\text{-}31a)$$

(the overlap between d_A and p is of sign opposite to that between d_A and σ).

In compounds of type (7-34), antiferromagnetism is observed[45] despite the large (5.6 Å) Cu—Cu distance. This is due to a large "through-bond" interaction[48] of the metal orbitals through the in-plane p orbitals of sulfur, ensuring a significant (0.6 eV) $(\epsilon'_- - \epsilon'_+)$ energy gap and a predominant antiferromagnetic second term in (7-30). The observed value of ΔE_{3-1} is 594 cm^{-1}.*

When the two metallic centers are different, the two orbitals d_A and d_B may be symmetry–orthogonal, as in (7-35) and illustrated in Fig. 7-18. Here S'_{AB} vanishes and the coupling becomes purely ferromagnetic ($\Delta E_{3-1} = -118$ cm^{-1})*. The understanding of such heteronuclear systems requires, however, careful consideration of an additional factor, the local energy difference between the two metal orbitals, which can eventually restore antiferromagnetism (pairing of electrons in the lower d orbital).

Finally, in copper oxalate (7-36) the observed antiferromagnetic splitting ($\Delta E_{3-1} = 291$ cm^{-1})* has been analyzed by molecular orbital theory and shown to be consistent with the proposed ribbon geometry with 5.14 Å Cu—Cu distances. Introduction of out-of-plane axial NH_3 ligands on the upper atoms modifies the metal orbitals (which now have less spread in the molecular plane) to such an extent that the overlap with the bridging oxygens falls to a very small value. The observed triplet-singlet splitting[47] drops concomitantly to 15 cm^{-1}!

7.9 Spin-Orbit Coupling: General Features

Spin-orbit coupling is the main physical factor that allows a molecule to proceed from triplet state to singlet state and conversely. The process may occur as intersystem crossing between spectroscopic states,[49,50] as collision-induced triplet to singlet crossing for reaction intermediates in the presence of inert gases,[51] or as triplet to singlet interconversion of organic diradicals. We are concerned here with

---

*The experimental J value given by certain authors[45–47] is *exactly* the singlet-triplet splitting, and not *half* this splitting as required by (7-9) and (7-28).

spin angular momentum

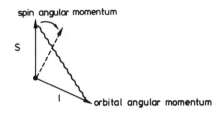

S

orbital angular momentum

FIGURE 7-19. Schematic description of spin-orbit coupling. The spin moment is pulled around by its interaction with the orbital angular momentum.

this last feature,[52,53] which plays an important role in a number of chemical reaction mechanisms.

The spin-orbit coupling Hamiltonian operator, in an approximation that neglects small two-electron effects such as coupling of an electron spin with the orbits of other electrons, is written as a sum of one-electron terms

$$\mathcal{H}_{\text{spin-orbit}} = \frac{e^2}{2m^2c^2} \sum_{\substack{\text{nuclei} \\ i}} \sum_{\substack{\text{electrons} \\ k}} \frac{Z_i}{r_{ik}^3} \boldsymbol{\ell}_k \cdot \mathbf{S}_k \qquad (7\text{-}38)$$

Each term involves the scalar product between the orbital angular momentum $\boldsymbol{\ell}_k$ of electron k and the spin angular momentum $\mathbf{S}_k$ of the same electron. The nuclear charge Z_i of nucleus i and the electron-nuclear distance r_{ik} are also involved.

Physically, as shown nicely by Kauzmann (Ref. 1, p. 349), (7-38) can be interpreted in the following manner. The magnetic moment associated with the orbital motion of the electron interacts magnetically with the magnetic moment associated with the electron's spin. This interaction, which operates as if there were a spring between the two moments, allows the orbital moment to *pull the spin moment around* (Fig. 7-19). The effect is particularly important in heavy atoms (factor Z_i) because orbital motion of the electron around the nucleus is equivalent, when we are concerned with what is happening at the position of the electron, to a motion of the nucleus "around" the electron.

Spin-orbit coupling is thus capable of flipping an α spin over into a β spin, thereby creating a transition between a singlet state and a triplet state. A closer inspection shows that no permanent orbital angular momentum is required in either state; the *instantaneous* orbital angular momentum change accompanying the spin flip suffices to induce the process. It is sufficient that the electron whose spin is flipped move out of its initial atomic or molecular orbital and move into a new orbital, so as to create a fleeting orbital angular momentum change. The ideal situation[52] is the electronic transition between a p orbital along one axis and a p orbital along an orthogonal axis (Fig. 7-20). The consequences of this prerequisite in diradicals are as follows:

1. One of the two diradical electrons, which flips its spin, must also switch into a different orbital. If we are dealing with p orbitals, the new p orbital must be orthogonal to the old one.

2. In a two-orbital system, such as a diradical with two available atomic or molecular orbitals, since the triplet has one electron in each orbital, the

FIGURE 7-20.   Ideal situation for effective intersystem crossing. The electronic transition that accompanies the spin flip creates orbital angular momentum.

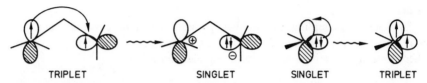

TRIPLET   SINGLET   SINGLET   TRIPLET

FIGURE 7-21.   Favorable intersystem crossing cases in diradicals with two available orbitals, showing the ionic or zwitterionic character of the singlet state: *(left)* trimethylene (see Section 3.6); *(right)* methylene (see Section 3.2).

singlet state generated by the electron flip-plus-switch operation must have some ionic or zwitterionic character (examples are shown in Fig. 7-21). Methylene (Fig. 7-21 RHS) turns out to be a very favorable case in this respect. The matrix element of interaction between singlet and triplet is ~ 38 cm^{-1}. Yet we know that the intersystem crossing is not very fast since singlet methylene and triplet methylene have distinct chemistries.[54]

A final, important condition for intersystem crossing is a perfect or nearly perfect degeneracy of the two interconverting states. A singlet with too much ionic character will "par force" be more stable than its purely covalent triplet counterpart, with ineffective intersystem crossing. The energy fit must be valid not only for the electronic states, but also for the vibrational levels since spin-orbit coupling is of the order of only a few cm^{-1}. This "vibronic" degeneracy is sometimes brought about by a foreign gas, or by side chains that provide a dense manifold of vibrational levels.

7.10   Spin-Orbit Coupling: Detailed Effects

It is illuminating to perform the very first quantitative step in the evaluation of the matrix elements of $\mathcal{H}_{\text{spin-orbit}}$ (7-38) between a singlet diradical

$$S_0 = \lambda(^1a^2) + \nu(^1ab) + \mu(^1b^2) \tag{3-9}$$

and the triplet built on the same open-shell orbitals

$$T_0 = {}^3ab \tag{3-6}$$

The matrix elements for the three singlet terms are respectively [in units of $Ze^2/(2m^2c^2)$]

$$\left.\begin{array}{c} \lambda\,\ell_{ab} \times \text{spin term} \\[4pt] \nu \times 0 \\[4pt] \mu\,\ell_{ab} \times \text{spin term} \end{array}\right\}, \quad \ell_{ab} = \int a\frac{\ell}{r^3}\,b\,d\tau \qquad (7\text{-}39)$$

where the weak overlap integral S_{ab} has been neglected and the spin term depends on the nature of the triplet spin component $(\alpha\alpha,\ \beta\beta,\ \alpha\beta + \beta\alpha)^*$ Hence the key to spin-orbit coupling is the size of the integrals, over the two odd orbitals a and b, of the three Cartesian components of the angular momentum operator[52,55,56]

$$\ell_{ab}^x = \int a\,\frac{\ell_x}{r^3}\,b\,d\tau$$

$$\ell_{ab}^y = \int a\,\frac{\ell_y}{r^3}\,b\,d\tau \qquad (7\text{-}40)$$

$$\ell_{ab}^z = \int a\,\frac{\ell_z}{r^3}\,b\,d\tau$$

This result has been cleverly exploited by Shaik[55,56] in a search for preferred orbital orientations and preferred nuclear motions inducing "spin inversion" or intersystem crossing in face-on olefin plus olefin cycloadditions (Section 4.2) and in Diels-Alder reactions (Section 4.3)

First, if orbitals a and b are on two different centers, the components ℓ_{ab} have maximum value when a and b are mutually perpendicular (see Section 7.9 for a qualitative discussion of this required orthogonality). Consider then a triplet photoreaction in which a donor and a triplet-sensitized acceptor approach each other to form an encounter complex, followed possibly by a triplet exciplex. As this exciplex goes on to singlet product, spin inversion must take place. Its efficiency will depend on the size of matrix elements of type (7-40). Now, in the cycloaddition of two double bonds in a face-on manner, the orbitals are all oriented in parallel fashion (Fig. 7-22). Spin-orbit coupling vanishes unless *promoting* motions can be

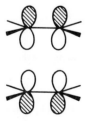

FIGURE 7-22.   Initial orientation of orbitals in the face-on cycloaddition of two olefins.

*Naturally the purely covalent part of the singlet wave function gives a zero contribution (see Section 7.9).

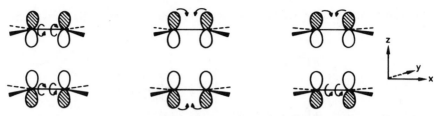

FIGURE 7-23. Promoting motions for D_{2h} face-on olefin cycloaddition for the three different triplet sublevels.[55] From left to right: "bis intramolecular disrotation," "bis pyramidalization" and combined "intramolecular pyramidalization and conrotation."

found that create mutually perpendicular atomic orbitals and optimize the spin-orbit coupling matrix elements. Since each triplet sublevel T_{0x}, T_{0y}, and T_{0z} has different total (spatial plus spin) symmetry and gives different spin terms in (7-39), each will be promoted, in its intersystem crossing potentiality, by different motions. Using the D_{2h} point group for the addends, Shaik finds[55] the promoting motions to be (1) a "bis intramolecular disrotation" for the sublevel T_{0x} (2) a "bis pyramidalization" for the sublevel T_{0y}, and (3) a combined "intramolecular pyramidalization and conrotation" for sublevel T_{0y}. These motions are represented schematically in Fig. 7-23.

Although all three motions ensure nonvanishing ℓ_{ab} values, their relative efficiency can be tested only by a full evaluation of these integrals. In this manner the "bis intramolecular disrotation" mechanism, which reaches a maximum for a 45° rotation by all four orbitals, takes place preferentially in a "loose" geometry—where it is governed by a large intramolecular term (the intermolecular term has the opposite sign). The "bis pyramidalization" mechanism, on the other hand, which reaches a maximum when all four initially trigonal centers reach the tetrahedral angle, takes place preferentially in a "tight" geometry—where intramolecular and intermolecular contributions reinforce each other.

Each spin inversion mechanism has its own stereochemical consequences. Hence stereospecificity in chemical reactions need not arise exclusively from the electronic features of singlet reactions; it may also follow from the spin-orbit coupling properties of reactions in triplet states.

References

1. W. Kauzmann, *Quantum Chemistry, An Introduction* (Academic, New York, 1957), p. 314 and Fig. 9-10.
2. C. A. Coulson, *Valence*, 2nd ed. (Oxford University Press, Oxford, 1961), p. 174.
3. F. Hund in *Linienspektren und periodisches System der Elemente* (Springer, Berlin, 1927), p. 124.
4. J. P. Colpa and M. F. J. Islip, *Molec. Phys.* **25**, 701 (1973).
5. J. P. Colpa and R. E. Brown, *Molec. Phys.* **26**, 1453 (1973).
6. J. P. Colpa, *Molec. Phys.* **28**, 581 (1974); J. P. Colpa, A. J. Thakkar, V. H. Smith, and P. Randle, *Molec. Phys.* **29**, 1861 (1975).
7. J. Katriel and R. Pauncz, *Adv. Quantum Chem.* **10**, 143 (1977).
8. H. M. McConnell, *J. Chem. Phys.* **24**, 764 (1956).
9. R. Bersohn, *J. Chem. Phys.* **24**, 1066 (1956).

10. S. I. Weissman, *J. Chem. Phys.* **25,** 890 (1956).
11. J. A. Pople and R. K. Nesbet, *J. Chem. Phys.* **22,** 571 (1954).
12. R. N. Zare, *J. Chem. Phys.* **45,** 1966 (1966).
13. R. J. Buenker and S. D. Peyrimhoff, *Chem. Phys.* **9,** 75 (1976).
14. W. T. Borden, *J. Am. Chem. Soc.* **97,** 5968 (1975).
15. H. Kollmar and V. Staemmler, *Theoret. Chim. Acta (Berl.)* **48,** 223 (1978).
16. H. Kollmar and V. Staemmler, *J. Am. Chem. Soc.* **99,** 3583 (1977).
17. W. T. Borden and E. R. Davidson, *Ann. Rev. Phys. Chem.* **30,** 125 (1979).
18. P. O. Löwdin, *Phys. Rev.* **97,** 1509 (1955); R. Pauncz, J. de Heer, and P. O. Löwdin, *J. Chem. Phys.* **36,** 2247 (1962).
19. D. B. Abraham and A. D. McLachlan, *Molec. Phys.* **12,** 319 (1967).
20. K. Yamaguchi, *Chem. Phys. Lett.* **30,** 288 (1975).
21. K. Yamaguchi, *Chem. Phys. Lett.* **28,** 93 (1974).
22. K. Yamaguchi and T. Fueno, *Chem. Phys. Lett.* **38,** 47 (1976); K. Yamaguchi and H. Fukutome, *Progr. Theoret. Phys.* **54,** 1599 (1975).
23. T. Nagamiya, *Solid State Phys.* **20,** 305 (1968).
24. B. Johansson and K. K. Berggren, *Phys. Rev.* **181,** 855 (1969).
25. A. A. Ovchinnikov, I. I. Ukrainski, and G. V. Kventsel, *Soviet Phys. Usp.* **15,** 575 (1973).
26. P. O. Löwdin, *Adv. Chem. Phys.* **2,** 207 (1957).
27. K. Yamaguchi and T. Fueno, *Chem. Phys. Lett.* **38,** 52 (1976).
28. W. Heisenberg, *Z. Physik* **38,** 411 (1926).
29. P. A. M. Dirac, *Proc. Roy. Soc.* **A123,** 714 (1929).
30. J. H. Van Vleck, *The Theory of Electric and Magnetic Susceptibilities* (Oxford University Press, London, 1932); *Phys. Rev.* **45,** 405 (1934).
31. G. R. Tsaparlis and S. F. A. Kettle, *Theoret. Chim. Acta (Berl.)* **45,** 95 (1977).
32. O. Kahn and B. Briat, *J. Chem. Soc., Faraday Transact. II* **72,** 268 (1976).
33. O. Kahn, *J. Chem. Soc., Faraday Transact. II* **71,** 862 (1975).
34. K. Yamaguchi, Y. Yoshioka, and T. Fueno, *Chem. Phys.* **20,** 171 (1977).
35. V. Bonacic-Koutecky, J. Koutecky, and L. Salem, *J. Am. Chem. Soc.* **99,** 842 (1977).
36. K. Yamaguchi, *Chem. Phys. Lett.* **34,** 434 (1975).
37. H. A. Kramers, *Physica* **1,** 182 (1934).
38. G. van Kalkeren, W. W. Schmidt, and R. Block, *Physica* **97B,** 315 (1979).
39. P. W. Anderson in *Solid State Physics*, Vol. 14, edited by F. Seitz and D. Turnbull (Academic, New York, 1963), p. 99.
40. L. Jansen, *Angew. Chem. Internatl. Ed.* **16,** 294 (1977).
41. P. J. Hay, J. C. Thibeault, and R. Hoffmann, *J. Am. Chem. Soc.* **97,** 4884 (1975).
42. O. Kahn and M. F. Charlot in *Quantum Theory of Chemical Reactions,* Vol. II, edited by R. Daudel, A. Pullman, L. Salem, and A. Veillard (Reidel, Dordrecht, 1981), p. 215.
43. V. H. Crawford, H. W. Richardson, J. R. Wasson, D. J. Hodgson, and W. E. Hatfield, *Inorg. Chem.* **15,** 2107 (1976)
44. M. F. Charlot, S. Jeannin, Y. Jeannin, O. Kahn, J. Lucrèce-Abaul, and J. Martin-Frere, *Inorg. Chem.* **18,** 1675 (1979).
45. J. J. Girerd, S. Jeannin, Y. Jeannin, and O. Kahn, *Inorg. Chem.* **17,** 3034 (1978).
46. O. Kahn, P. Tola, J. Galy, and H. Coudanne, *J. Am. Chem. Soc.* **100,** 3931 (1978).
47. J. J. Girerd, O. Kahn, and M. Verdaguer, *Inorg. Chem.* **19,** 274 (1980).
48. R. Hoffmann, *Acc. Chem. Res.* **4,** 1 (1971).
49. S. P. McGlynn, T. Azumi, and M. Kinoshita, *Molecular Spectroscopy of the Triplet State* (Prentice-Hall, Englewood Cliffs, N.J., 1969).
50. M. El-Sayed, *Pure Appl. Chem.* **24,** 475 (1970).
51. T. W. Eder and R. W. Carr, *J. Chem. Phys.* **53,** 2258 (1970).
52. L. Salem and C. Rowland, *Angew. Chem. Internatl. Ed.* **11,** 92 (1972).
53. L. Salem, *Pure Appl. Chem.* **33,** 317 (1973).
54. J. A. Bell, *Progr. Phys. Org. Chem.* **2,** 1 (1964).
55. S. S. Shaik, *J. Am. Chem. Soc.* **101,** 3184 (1979).
56. S. S. Shaik and N. D. Epiotis, *J. Am. Chem. Soc.* **102,** 122 (1980).

8

Solvent Properties and the Different Models for Studying Solvent Effects

There is a great difference between studying the behavior of isolated molecules and studying their reactions in solution. The details of the way in which surrounding solvent molecules modify the approach of two reacting molecules remain a mystery. General qualitative rules will eventually be found to account for the specific electronic interactions between solute and solvent, and to describe the dynamical effects due to the motion of the medium molecules relative to the reacting pair.

At the present time, definite trends are already known that relate solvent polarity, donor-acceptor characteristics, or polarizability to reaction rates and spectral shifts. Also, theoretical efforts to calculate solvation energies and to determine the exact configuration of solvation shells have recently multiplied. Models in which the solvent is treated as a continuum compete with those that rely on the precise position of the closest solvent molecules. With growing computing power, more will be discovered on the detailed behavior of the solvent molecules and the relation of this behavior to observable macroscopic properties.

8.1 Solvent Effects on Reaction Rates: Solvation as a Reaction Coordinate

Certain major solvent effects have long been understood in organic chemistry:

Protic solvents can solvate anions by hydrogen bonding whereas aprotic solvents cannot do so.[1]

Polar solvents increase the rates of reactions proceeding through charged or dipolar transition states. An example is the addition of bromine to olefins[2]:

$$\diagup\!\!\!\diagdown \!\!\!< + \ Br_2 \longrightarrow \diagup\!\!\!\!\diagdown\!\!\!\!\diagdown \ \text{(bromonium ion), } Br^{\ominus} \qquad (8\text{-}1)$$

Nonpolar solvents do not have this effect. Table 8-1 shows a classification of solvents according to their protic character and polarity.[3]

The donor-acceptor characteristics of a solvent, through lone pairs occupying high-energy orbitals or through holes in low-lying vacant orbitals, are distinct

214

TABLE 8-1

Classification of a Few Solvents According to Protic Character and Polarity[a]

Protic	ϵ	Aprotic	ϵ
		Weakly Polar	
Acetic acid	6.15	*n*-Hexane	1.89
Phenol	9.78	Carbon tetrachloride	2.24
Ammonia	16.9	Tetrahydrofuran	7.58
		Strongly Polar	
Ethanol	24.5	Acetone	20.7
Methanol	33.6	Nitromethane	35.9
Formic acid	58.5	Acetonitrile	37.5
Water	80.4	Dimethylsulfoxide	46.7

Source: Ref. 3.

[a]Dielectric constants (at 20°C if known) are also given.

from its polarity characteristics (molecular dipole moment, dielectric constant). Section 6.4 emphasized this point for acids and bases.

The theoretical description of the change between a reaction in the gas phase and a reaction in solution has many levels of sophistication. The most elaborate account would add, to the nuclear coordinates of the reaction system proper, the coordinates for translation and rotation of all the surrounding solvent molecules—and even, properly, their internal valence coordinates. We return to the principles of such a description in Section 8.8. The crudest account, however, uses solvation as a *single* additional nuclear coordinate. This "coordinate" generally represents some average, ill-described, distance of approach between the reaction system and the solvent molecules. In practice, initial values of this coordinate correspond to very weak solvation, with poor orientation of the solvent by the reacting solute, whereas large values correspond to a very dense solvent surrounding. In between there are a series of related solvent configurations that are optimal for the reacting solute in its initial geometry, for the transition structure and for its final geometry.

Figure 8-1 illustrates the corresponding three-dimensional energy graph for a weakly polar reactant going to a weakly polar product through a strongly dipolar transition state, such as occurs in (8-1). In the gas phase the reaction proceeds from point R_1 (reactant), through the transition state TS_1, to product P_1. The energy profile is assumed to have a large hump at the geometry corresponding to the transition state.

Now we expect the energy of reactant and product to be somewhat sensitive to solvation, but less so than the transition state. Hence if we introduce a polar solvent and *fix* it in such a manner as to optimize the stabilizing interactions between solvent molecules and transition structure, the profile flattens out. The reaction, from R_2 (reactant), to TS_2 and P_2 is far less costly in energy. If we approach the solvent even more, the entire profile rises because steric interactions with solvent molecules that are too close are energetically unfavorable. Starting at point R_3 along the solvation coordinate, we progress to TS_3 and P_3. The overall construction

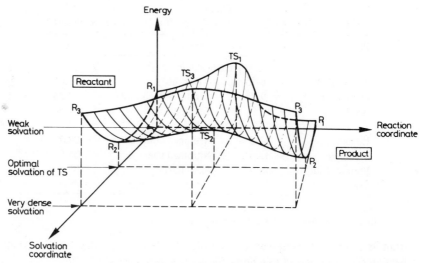

FIGURE 8-1.   Use of a single reaction coordinate to describe polar solvation of a dipolar transition state.

shows an energy sheet with a favorable pathway along the valley through the optimally solvated transition state. This simplified description does not account for changes in solvent position that *accompany* the molecular reaction. The true pathway must involve a mixture of solvent motion and reactive solute motion.

The Hughes-Ingold rules[4,5] constitute a direct application of Fig. 8-1 to aliphatic nucleophilic substitution and elimination reactions. They state[5] that an increase in solvent polarity—in the sense of solvent power to solvate charges—increases the rates of those reactions in which (1) the *charge separation is greater* (larger effective dipole moment) or (2) the *overall charge is more compact* (larger effective net charge in a given radius) in the activated complex than in the initial reactant molecules. The situation is again described by Fig. 8-1; although the reactant R_1 is stabilized (to R_2) by solvation, it is not stabilized *as much* as the transition structures TS_1; hence the larger rate along path 2. Such an increase in charge separation occurs in reactions such as[5]:

$$R—X \longrightarrow R^{\delta+} \cdots X^{\delta-} \qquad (S_N1)$$

$$Y + R—X \longrightarrow Y^{\delta+} \cdots R \cdots X^{\delta-} \qquad (S_N2)$$

$$H—\overset{|}{\underset{|}{C}}—\overset{|}{\underset{|}{C}}—X \longrightarrow H—\overset{|}{\underset{|}{C}}—\overset{|}{\underset{|}{C}}^{\delta+} \cdots X^{\delta-} \qquad (E1)$$

$$Y: + H—\overset{|}{\underset{|}{C}}—\overset{|}{\underset{|}{C}}—X \longrightarrow Y^{\delta+} \cdots H \cdots \overset{|}{C} \overset{...}{=} \overset{|}{C} \cdots X^{\delta-} \qquad (E2)$$

$$(8\text{-}2)$$

The opposite effect, a rate decrease, is observed (1) if the charge separation is smaller in the transition state or (2) if the charge in the transition state is less compact, leading to weaker solvation by a polar solvent. Less compact charges occur in transition states for

$$R\!-\!X^+ \longrightarrow R^{\delta+} \cdots X^{\delta+} \qquad\qquad (S_N1)$$

$$Y^- + R\!-\!X \longrightarrow Y^{\delta-} \cdots R \cdots X^{\delta-} \qquad (S_N2)$$

$$\text{H}\!-\!\overset{|}{\underset{|}{\text{C}}}\!-\!\overset{|}{\underset{|}{\text{C}}}\!-\!\text{X}^+ \longrightarrow \text{H}\!-\!\overset{|}{\underset{|}{\text{C}}}\!-\!\overset{|}{\underset{|}{\text{C}}}{}^{\delta+} \cdots X^{\delta+} \qquad (E1)$$

$$Y^{\overline{\cdot}} + \text{H}\!-\!\overset{|}{\underset{|}{\text{C}}}\!-\!\overset{|}{\underset{|}{\text{C}}}\!-\!\text{X} \longrightarrow Y^{\delta-} \cdots \text{H} \cdots \overset{|}{\text{C}}\!-\!\overset{|}{\text{C}} \cdots X^{\delta-} \qquad (E2)$$

$$(8\text{-}3)$$

For such reactions, the potential-energy sheet in Fig. 8-1 would have its valley displaced toward R_1, for weak values of the solvation ''coordinate,'' and TS_1 would lie lower than TS_2.

8.2   Solvent Effects on Absorption and Emission Spectra*

Although ultraviolet spectra are not strictly a topic of this book, a brief study of solvent-induced spectral shifts serves as a good introduction to solvent effects on the energies of molecular states. Let us first summarize the different types of forces involved:

	Solvent	
Solute	Nonpolar	Polar
Nonpolar	Dispersion	Dispersion Polarization
Polar	Dispersion Polarization	Dispersion Polarization Electrostatic

$$(8\text{-}4)$$

The dispersion forces are the weak London-van der Waals forces that arise between instantaneous dipoles on the two molecules, solute and solvent. The polarization forces result from the Coulomb interaction between a permanent moment on a polar molecule and the induced moment in the partner (whether polar or nonpolar). Finally, electrostatic forces are the classical Coulomb forces between permanent moments (dipoles, usually) on polar molecules.

Consider first a single solute molecule (dipole moment $\boldsymbol{\mu}$) surrounded by a great number of nonpolar but polarizable solvent molecules. As a result of polarization

---

*For theoretical studies of solvent effects on ultraviolet spectra, see Refs. 6–14.

forces, the dipole induces small moments in the solvent, which in turn, lead to an "electronic" reaction field at the center of the solute molecule[15]:

$$\mathbf{E} = -2\left(\frac{n^2 - 1}{2n^2 + 1}\right)\frac{\boldsymbol{\mu}}{a^3} \tag{8-5}$$

In (8-5) n is the refractive index of the nonpolar medium and a the radius of the "cavity" assumed to surround the solute†.

Consider next the case where the solvent molecules are themselves polar, with a macroscopic dielectric constant ϵ. The electronic reaction field is still present, but there is superimposed a more important reaction field $\mathbf{E}'$ due to the reorientation of the solvent molecules in the field of the solute:

$$\mathbf{E}' = -2\left(\frac{\epsilon - 1}{2\epsilon + 1} - \frac{n^2 - 1}{2n^2 + 1}\right)\frac{\boldsymbol{\mu}}{a^3} \tag{8-6}$$

The total reaction field is the sum of this "orientational" reaction field and of the electronic reaction field†:

$$\mathbf{E} + \mathbf{E}' = -2\left(\frac{\epsilon - 1}{2\epsilon + 1}\right)\frac{\boldsymbol{\mu}}{a^3} \tag{8-7}$$

(these expressions are strictly correct only if the solute molecule is non-polarizable[13]).

Let us now consider an electronic transition that carries a polar molecule from its ground state ($\boldsymbol{\mu}_g$) to its excited state ($\boldsymbol{\mu}_e$) in a polar solvent medium. In the vertically excited state all the nuclear coordinates are identical with those in the ground state, in conformity with the Franck-Condon principle.[16,17] The electronic transition is completed before the solvent molecules have time to reorient. Only the reaction field $\mathbf{E}$ due to induced polarization of solvent has time to adapt to the new size and orientation $\boldsymbol{\mu}_e$ of the solute dipole moment as shown in Fig. 8-2. Any "orientational" reaction field, however, has the same size and direction $\mathbf{E}'_e$ in the excited state as in the ground state $\mathbf{E}'_g$. Hence, in a polar solvent, the orientational stabilization of the ground state is accompanied by an orientational destabilization (or lesser stabilization)* of the excited state. Compared with the case of a nonpolar solvent, in which $\mathbf{E}' = 0$, there is a displacement to higher energies of the transition energy (Fig. 8-3). Thus the *blue* shift in the *absorption* spectra of dipolar solutes as the solvent is changed from nonpolar to polar.

If time is allowed for the solvent to relax around the excited polar molecule, the excited level is stabilized. Hence *fluorescence* in a polar solvent is *red*–shifted

---

*Whether the excited level actually rises or descends depends on the orientation of the excited dipole relative to ground dipole. The excited state, relative to the nonpolar case, may be slightly stabilized if $\boldsymbol{\mu}_e$ is not too poorly oriented relative to the prearranged ground polar solvent. But in either case the ground state will be stabilized even more; hence a blue shift.

†Owing to our convention for dipoles (oriented from positive end toward negative end) the reaction field is *opposed* to the solute dipole [minus sign in (8-5) and (8-7)] while *stabilizing* it: the field points its negative end at the positive end of the dipole; see Fig. 8-2, lower part. The author is grateful to J. L. Rivail for a discussion of reaction fields.

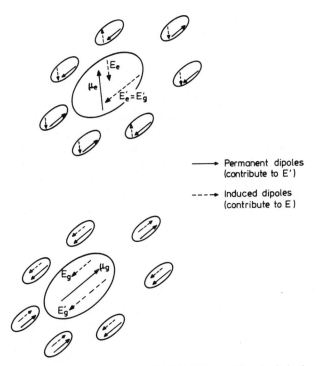

FIGURE 8-2.   Polar solute in a polar solvent: reaction fields **E** due to induced polarization of the solvent molecules and **E′** due to reorientation of the solvent molecules. *Above:* excited solute molecule. *Below:* ground solute molecule (both reaction fields stabilize the inducing moment **μ**$_g$). In the Franck-Condon vertically excited state the orientation field $\mathbf{E}'_e = \mathbf{E}'_g$.

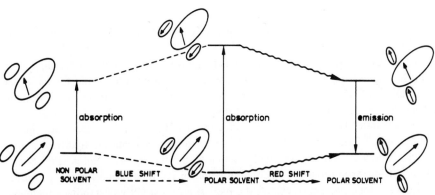

FIGURE 8-3.   Spectral shifts accompanying *(a)* a solvent change (absorption); *(b)* the change from absorption to fluorescence (polar solvent). The solvent is schematized by two small molecules around a large solute molecule.

219

relative to its absorption, as illustrated in Fig. 8-3, because of this stabilization and the corresponding destabilization of the ground state to which decay now occurs in a Franck-Condon transition without time for orientational relaxation of the solvent.

Another interesting comparison concerns the spectral shifts in a series of solvents with different polarities. If μ_g and μ_e have nearly the same orientation, the relative stabilization of ground and excited state when increasing the solvent polarity depends on whether $\mu_g < \mu_e$ (red shift) or $\mu_g > \mu_e$ (blue shift).[18] Another case of red shifting occurs in acrolein-H_2O complexes (relative to the monomers), where the hydrogen bonds between solute and solvent are stabilized in the $\pi \to \pi^*$ excited states due to an increase in the carbonyl negative π charge on the excitation.[19] On the contrary, passage from nonpolar or aprotic polar solution to a protic polar solution causes the n,π^* band of carbonyl compounds such as benzophenone or acetone to be blue–shifted.[20] Here desolvation in the excited state is brought about by loss of σ charge on oxygen on excitation.

8.3 Reversal of Stability Due to Solvation

A striking example of solvent effect on the relative stability of a series of related molecules is given by the acidity of alcohols and of amines. In solution, the relative acidities

$$\text{(solution)} \quad OH_2 \text{ (stronger)} > \text{MeOH} > \text{EtOH} > t\text{-BuOH (weaker)} \quad (8\text{-}8)$$

had long been ascribed to the donor characteristics of the methyl group that would stabilize cations and destabilize anions. Hence HO^- would be more stable than MeO^-, itself more stable than EtO^-, and so on. However, in the gas phase, the order is reversed[21]:

$$\text{(gas phase)} \quad t\text{-BuOH (stronger)} > \text{EtOH} > \text{MeOH} > OH_2 \text{ (weaker)} \quad (8\text{-}9)$$

Since the gas-phase situation corresponds to the isolated molecules, it would appear that the *solvent,* rather than the methyl groups, must be responsible for the order in solution and that concurrently, the role of the methyl substituent in the isolated molecule may not always be that of a donor.

This is indeed the hypothesis first advanced by Hudson, Eisenstein, and Nguyen Trong Anh.[22] They start from the states of neutral water and the hydroxyl anion, to which they compare respectively methanol and the methoxy anion (Fig. 8-4).* By a four-electron repulsive interaction between its filled pseudo-π (π_{CH_3}) orbital and the $p\pi$ lone pair on oxygen,† the methyl group slightly destabilizes methanol relative to water. In the methoxy ion, stabilizing two-electron interactions occur

---

*The level scheme shown here concerns only the two most labile electrons that carry the negative charge in the anion. Thermodynamically, for the overall energy, the CH_3OH,CH_3O^- pair lies far below the OH_2,OH^- pair of levels.

†For group orbitals in molecules such as methanol and water, see Ref. 23.

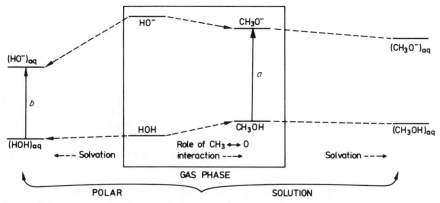

FIGURE 8-4.   Relative acidities of methanol and water: *(a)* in the gas phase, methanol is more acidic; *(b)* in polar solution water is more acidic.

between the empty pseudo–π $(\pi^*_{CH_3})$ and $(\pi'^*_{CH_3})$ orbitals and the two oxygen pairs $(p\pi$, which has been raised in energy, and the newly formed $p\pi')$; hence the methoxy ion is slightly stabilized relative to the hydroxyl ion.

If we now place the molecules in a polar protic solution, both anions are, of course, stabilized by solvation. The smaller anion tends to be better solvated by the electrostatic forces involved. Furthermore, hydrogen bonding with the solvent stabilizes the oxygen lone pair orbitals and decreases their availability for the two-electron effect that caused relaxation of the isolated methyl-substituted ions. These two effects both work in the same direction, and stabilize the smaller, less substituted ions relative to the larger, substituted ions. Consequently, water is more acid than methanol in solution.

This example illustrates the versatile nature of the methyl substituent.[24] For instance, if methyl substitution increases the acidity of water, it decreases the acidity of acetylene.[25] This time, the major role is played by the two-electron interaction between π_{CH_3} and the antibonding orbitals of the triple bond. However, the $\pi^*_{CH_3} \leftrightarrow p\pi$ lone-pair interaction that predominated in methanol is again responsible for the higher acidity of methylamine relative to ammonia[26] and indicates the existence of "anionic hyperconjugation" resonance structures of the type

$$H_3C{-}X \leftrightarrow H_2\overset{\ominus}{C}{=}X \leftrightarrow \cdots \quad (X = NH, O, \text{ etc.}) \qquad (8\text{-}10)$$

The order of nucleophilicity of small anions is also reversed by protic solvation:

$$\text{(gas phase or aprotic solvent) } I^- \text{ (stronger)} > Br^- > Cl^- > F^- \text{ (weaker)}$$
$$\text{(polar protic solvent) } F^- \text{ (stronger)} > Cl^- > Br^- > I^- \text{ (weaker)} \qquad (8\text{-}11)$$

Here the phenomenon is related to the position of the highest occupied molecular orbital (HOMO).[27] In a frontier orbital interaction picture (Sections 6.3 and 6.4) this orbital is deeply involved in the primary step of the nucleophilic substitution

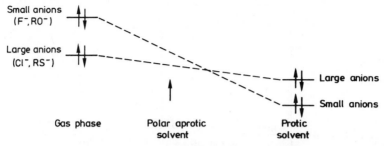

FIGURE 8-5.   Position of HOMO for anions of different size.

reaction. In the gas phase the *smaller* anions have the higher and more available HOMO (due to a larger interelectronic repulsion that pushes up all the occupied orbitals) and the more nucleophilic character. However, when surrounded by a protic solvation shell, these same ions, which are more strongly involved in hydrogen bonding, see their HOMO become lower than that of the *larger* anions (Fig. 8-5). In aprotic solution conditions remain similar to the gas phase, although the differences in nucleophilicity are less acute.

Ions with different extents of charge delocalization can also be affected in a differential manner by solvation. For instance, the classical ethyl cation (**1**), calculated to be less stable than the bridged cation (**2**) by 29 kJ/mole, becomes more stable by 59 kJ/mole when both are solvated (by a model HCl solvent).[28] Indeed, the net charge is more concentrated, and the solvent stabilization larger, in the classical cation than in the nonclassical one.

$$\text{(8-12)}$$

8.4   The Solvated Electron

This section draws largely on the superb reviews by Dainton[29] and by Matheson.[30] The solvated electron—as distinct from the electron trapped in an amorphous solid or a glass—can be obtained by radiolysis of the appropriate solvent (water, ammonia, amines, ether, etc.), dissolution of metals in these same solvents, flash photolysis of inorganic anions or organic solutes such as aniline and phenol, and other sophisticated methods (photoemission, field emission from metals, etc.).

The theoretical models for treating the solvated electron and understanding its physical properties (ultraviolet spectrum, EPR spectrum, etc.) can range from the very crude to the highly sophisticated. The range of models serves as a convenient introduction to the general models for solvation of molecules (Sections 8.5–8.10). We restrict our discussion here to the localized electron as opposed to the delocalized entity that occurs in liquid rare gases or nonpolar hydrocarbons.

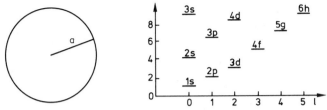

FIGURE 8-6. Energy levels [in units of $\pi^2/(2a^2)$] for a particle in a spherical box.[31]

The simplest model would be to imagine the electron as a particle in a spherical cavity defined by the surrounding shell of solvent molecules (Fig. 8-6). The wave functions and energy levels for such a problem are well known.* The lowest energy levels are shown in Fig. 8-6. The striking feature in the present system as opposed to the hydrogen–like atom, where functions with different ℓ values have the same energy, is the decrease in energy with increasing angular quantum number ℓ. If we identify the main absorption band of the solvated electron in water at 7150 Å with the $1s \rightarrow 2p$ transition of the particle in the cavity, and with energy $1.04 \times \pi^2/(2a^2)$, we obtain

$$a = 4.75 \text{ Å} \qquad (8\text{-}13)$$

for the cavity radius.

A second model would try, by using molecular orbitals, to allocate the solvated electron to an appropriate low-lying antibonding orbital of the solvent molecules. It might be assumed, for instance, that a particular geometric configuration of the first solvent shell of molecules stabilizes a combination of the lowest antibonding orbitals of each molecule. For the hydrated electron, the solvating water molecules probably have their hydrogen atom pairs pointing toward the electron (Fig. 8-7a).[32] Since the lowest antibonding orbital of each molecule is strongly concentrated on the hydrogen atoms (Fig. 8-7b),[23] each such orbital can interact with the corresponding members of the other solvent shell molecules. The formation of a very low lying orbital in this manner has, however, yet to be proven.

The next step performs an actual calculation† on *clusters* of the type $(H_2O)_4^-$, $(H_2O)_5^-$, $(H_2O)_6^-$. One tries to evaluate the stability of the anionic cluster relative to the corresponding neutral; hence the affinity of the cluster for the electron. The results fall in the range 1 eV[34] to 2.4 eV.[33] But such calculations, which by nature are restricted to the short-range interactions between the charge and the first solvent shell, neglect the important effect of the electron on the polarizable "continuous" solvent medium that surrounds this shell. True sophistication obtains when the

*See Ref. 31. There is an error in equation (B-13) of that reference, where $1/r$ should be replaced by $1/\sqrt{r}$.

†For a tetramer, the first calculation is due to Natori and Watanabe.[33] Many semiempirical calculations followed.[34] See Ref. 32 for a review.

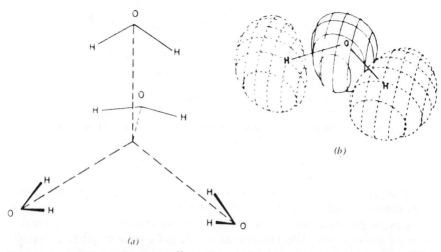

FIGURE 8-7.   *(a)* Dipole oriented model[32] for $(H_2O)_4^-$; *(b)* lowest antibonding orbital of water.[23] Part *(a)* reprinted with permission from J. D. Newton, *J. Phys. Chem.* **79**, 2795 (1975). Copyright 1975 by American Chemical Society.

electron and its solvent cluster and the polarizable surrounding continuum are all accounted for[32,35–37], as was first shown in the pioneering work of Jortner.

To illustrate such attempts, we outline here the method due to Newton,[32] which combines a detailed self-consistent field (SCF) molecular orbital approach at short distances with long-range polarization of the medium. Newton writes the total energy of the system as

$$E = E_0 + U \qquad (8\text{-}14)$$

where E_0 is the electronic energy of the excess charge plus that of a discrete cluster of molecules and U is the polarization energy of the dielectric continuum. This continuum is supposedly an infinite dielectric of volume V with a spherical cavity of radius a. The general expression for U is the free energy required to transfer the charge from vacuum to within the polarizable continuum of dielectric constant ϵ. Hence[38]

$$U = -\frac{1}{8}\left(1 - \frac{1}{\epsilon}\right)\int_V \mathbf{F}^2 \, d\tau \qquad (8\text{-}15)$$

where $\mathbf{F}$ is the field *produced* in the medium of dielectric constant ϵ by the charge:

$$\mathbf{F} = \int_{\text{all space}} \rho(r') \frac{|\mathbf{r} - \mathbf{r}'|}{|\mathbf{r} - \mathbf{r}'|^3} \, d\tau' \qquad (8\text{-}16)$$

[(the field $\mathbf{F}$ should not be confused with the reaction fields $\mathbf{E}$ and $\mathbf{E}'$ in (8-5) and (8-6), which also lead to expressions for U such as (8-32)]. Newton simplifies the double integral implicit in (8-15) by employing only the spherical component of $\rho(\mathbf{r})$—hence the disappearance of $1/(4\pi)$—and by truncating the Coulomb interaction term:

$$U = \frac{1}{2}\left(1 - \frac{1}{\epsilon}\right)\iint_{\text{all space}}\rho(\mathbf{r_1})\left[-\frac{1}{\max{(r_1, r_2, a)}}\right]\rho(\mathbf{r_2})d\tau_1\,d\tau_2 \quad (8\text{-}17)$$

where the function in brackets is called $g(r_1, r_2)$.

The wave function and the density are obtained by minimizing (8-14). This leads to a generalized Hartree-Fock equation

$$F'\psi = (F + f)\psi = \epsilon\psi \qquad (8\text{-}18)$$

[compare with (1-19)], where f represents the contribution from the polarized dielectric

$$f(r) = \left(1 - \frac{1}{\epsilon}\right)\int_{\text{all space}}\rho(\mathbf{r'})g\,(r, r')d\tau' \qquad (8\text{-}19)$$

to the trapping potential of the electron.

The calculations are carried out for various geometrical configurations of the discrete cluster, such as in Fig. 8-7a where the solvent dipoles point toward the cavity center, or where single hydroxyl bonds point directly at the center. For the model in Fig. 8-7a, an equilibrium cavity–center–to–oxygen distance of r_d = 2.65 Å is found that, together with an effective solvent molecule radius of 1.5 Å, yields

$$a = r_d + r_s \approx 4.15\ \text{Å} \qquad (8\text{-}20)$$

for the hydrated electron (see also Fig. 8-8). This result compares quite nicely with the very simple, particle-in-a-box model [see (8-13)].

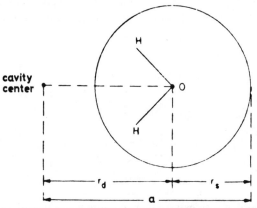

FIGURE 8-8. Definition of the various radii.[32] Reprinted with permission from J. D. Newton, *J. Phys. Chem.* **79**, 2795 (1975). Copyright 1975 by American Chemical Society.

In this calculation $(H_2O)_4^-$, in the dipole-oriented configuration, is found to be 1.62 eV more stable than $(H_2O)_4$, whereas the hydrated electron is calculated to have an excitation energy of 1.20 eV (experimental value 1.73 eV) and a vertical ionization energy of 2.51 eV.

8.5   Continuum Theories of Solvation: Virtual Charge Techniques

Three general approaches can be used to evaluate the effects of solvent on a reaction path. At the present time these approaches have been applied only to evaluate solvation energies of simple solutes, to estimate the effect of solvation on conformational energy differences, or again to map out the solvent surroundings of a solute. But the extension from simple geometries to full reaction paths should not be long to come.

Continuum theories (Fig. 8-9*a*) describe the solute as lying in a cavity in contact with a polarizable continuum representing the solvent. In semicontinuum theories (Fig. 8-9*b*) the first shell or first few shells of solvent molecules around the solute are treated in a detailed manner, as discrete entities. The remaining solvent molecules are still treated as a continuum. Newton's description of the solvated electron (Section 8.4) belongs to the "semicontinuum" family. Finally, the fully discrete theories (Fig. 8-9*c*) treat as many solvent molecules as possible in full quantum-mechanical detail, on the same footing as the solute molecule.

The continuum model itself has been adapted in two manners. Both methods put the solute molecule(s) in a cavity surrounded by the polarizable medium. But in one case the effect of the solvent is calculated by ascribing to the solute molecules effective charges that differ from the true net charges carried by their atoms. In the other case the interaction between the solute and its reaction field is included explicitly in the Schrödinger equation.

In 1967 Klopman[39] introduced the "solvaton" model, in which each atom of the solute molecule is surrounded by an imaginary particle, the "solvaton," whose charge is equal in magnitude but opposite to that of the atom to which it is attached. This solvaton is assumed to account for the oriented solvent distribution around the charged atom (Fig. 8-10). For an atom of net charge δq_A and radius r_A, its interaction with its solvaton of net charge $-\delta q_A$ is assumed to be

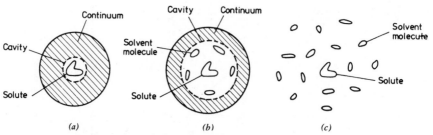

FIGURE 8-9.   Various models for treating solvation: *(a)* continuum; *(b)* semicontinuum; *(c)* discrete.

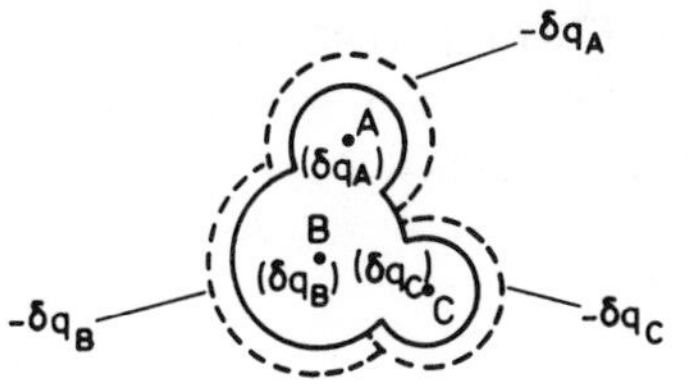

FIGURE 8-10. Klopman's solvatons.

$$U = -\frac{1}{2}\frac{(\delta q_A)^2}{r_A}\left(1 - \frac{1}{\epsilon}\right) \tag{8-21}$$

which is the correct Born solvation energy for a charge δq_A placed in a continuum[40] if the parameter ϵ is identified with the dielectric constant ϵ of the continuous solvent. However, there is little theoretical justification for writing an expression such as (8-21) for the interaction between two oppositely signed charges on the same center.

Hence Constanciel and Tapia[41,42] have refined this idea in their "virtual charge model," where, in the presence of a solute bearing a polarizing charge Q_0, the polarized solvent carries at the cavity surface a superficial "virtual" charge Q (Fig. 8-11)

$$Q = \alpha Q_0 \tag{8-22}$$

where the value of α remains to be determined. The virtual charge simulates the solvent and *both* its induced *and* orientational polarization by the solute. The total energy is

$$U = \frac{1}{2}\frac{Q_0^2}{a} + \frac{Q_0 Q}{a} + \frac{1}{2}\frac{Q^2}{a} \tag{8-23}$$

where a is the cavity radius. The first term is the self-energy of the solute charge, the third term is the self-energy of the solvent charge, and the middle term is the interaction between Q_0 and Q. Hence

$$U = U_0 + \frac{Q_0^2}{2a}\alpha(2 + \alpha) \tag{8-24}$$

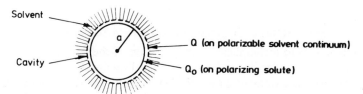

FIGURE 8-11. Polarizing charge Q_0 and virtual charge Q.

where U_0 is the energy of the solute in vacuum. Clearly, when the dielectric constant ϵ of the solvent is 1, α must vanish; however, for $\epsilon \to \infty$, $\alpha \to -1$ [see (8-22)]. Constanciel and Tapia thus choose the empirical function.

$$\alpha = - \left(1 - \frac{1}{\sqrt{\epsilon}}\right) \tag{8-25}$$

Equation (8-24) then gives a particle solvation energy

$$U - U_0 = - \left(1 - \frac{1}{\epsilon}\right) \frac{Q_0^2}{2a} \tag{8-26}$$

which is the correct Born expression. The improvement over the solvaton theory is that *both* the induced charge and the solvation energy depend reasonably on ϵ, whereas for the solvaton, α is arbitrarily set equal to -1 and equation (8-21) must be given as a "rule." However, the expression deduced from (8-22) and (8-25)

$$\frac{Q}{a} = - \left(1 - \frac{1}{\sqrt{\epsilon}}\right) \frac{Q_0}{a} \tag{8-27}$$

is slightly different from the correct Onsager reaction field[15]

$$\mathbf{E} + \mathbf{E}' = - \left(1 - \frac{1}{\epsilon}\right) \frac{Q_0}{a} \tag{8-28}$$

for a polarizing charge Q_0. It is indeed impossible to find a value of α that satisfies both (8-26) and (8-28).

The "virtual charge model" theory can easily be generalized to a polyatomic solute. Within the CNDO approximation (Section 1.9) one need only add algebraically to the approximate Fock matrix element $F_{\mu\mu}$ on center μ a term due to the induced virtual charge at that center. The net charge on each atom is decreased accordingly. The method has been applied to a pair of interacting water molecules.[43] It reveals nicely a double-minimum potential. One of the minima originates from the pair of neutral molecules, and the other is ascribed to the H_3O^+, OH^- ion-pair (Fig. 8-12). This second minimum disappears for small dielectric constants. Recently an improved version of the virtual charge model, the "image potential method," has been developed.[44]

8.6 Continuum Theories of Solvation: Reaction Field Techniques

In the reaction field techniques[45,46] the reaction of the solvent, as a polarized continuum*, to the polarizing solute is calculated rigorously—and the interaction energy between the two included in the Hamiltonian. The solute molecule is rep-

*This polarization includes both induced polarization and orientational polarization (see Section 8.2).

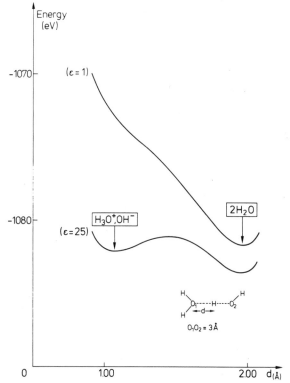

FIGURE 8-12.   Application of virtual charge model to a pair of water molecules.[43]

resented by n point charges q_i situated at points $\mathbf{r}_i$ inside a sphere of radius a (Fig. 8-13). The potential at a point P (defined by the vector $\mathbf{r}$) *inside* the sphere, as a function of the dielectric constant ϵ of the polarizable continuum *outside* the sphere, is given by Kirkwood[47]:

$$V(r) = \sum_i \left\{ \frac{q_i}{|\mathbf{r} - \mathbf{r}_i|} + \frac{q_i}{a} \sum_{\ell = 0}^{\infty} \frac{(\ell + 1)(1 - \epsilon)}{\epsilon(\ell + 1) + \ell} \left(\frac{r\, r_i}{a^2} \right)^{\ell} P_\ell(\cos \theta_i) \right\} \quad (8\text{-}29)$$

The first term is the classical potential in vacuum, and the second term is the potential created at P by the dielectric continuum polarized by the charge q_i.

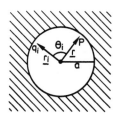

FIGURE 8-13.   Point charge q_i in a cavity embedded in the continuum.

The interaction energy between the solute molecule and the solvent medium is then given by

$$U = \frac{1}{2} \int \rho(\mathbf{r}) V(\mathbf{r}) d\tau \tag{8-30}$$

for a continuous density ρ [(8-30) can be related to (8-15) through Poisson's equation] or by

$$U = \frac{1}{2} \sum_i q_i V(\mathbf{r}_i) \tag{8-31}$$

for discrete charges. For a molecular system with nuclei of charge Z_α at $\mathbf{R}_\alpha$ and electrons at $\mathbf{r}_i$, (8-29) and (8-31) can be used to write the general expression

$$U = \sum_{\ell = 0}^{\infty} U_\ell \tag{8-32}$$

In (8-32) the ℓth component U_ℓ of the solvation energy appears as a product between components M_ℓ^m of electric moments of order 2_ℓ and components E_ℓ^m of the total reaction field with the *same* symmetry as the moments. We have

$$U_\ell = \frac{1}{2} \sum_{m = -\ell}^{\ell} E_\ell^m M_\ell^m \tag{8-33}$$

with

$$E_\ell^m = - \frac{(\ell + 1)(\epsilon - 1)}{\epsilon(\ell + 1) + \ell} \frac{1}{a^{2\ell+1}} M_\ell^m \tag{8-34}$$

(see for instance (8-7) for $\ell = 1$) and

$$M_\ell^m = - \sum_\alpha Z_\alpha S_\ell^m (\mathbf{R}_\alpha) + \sum_i S_\ell^m (\mathbf{r}_i) \tag{8-35}$$

$$S_\ell^m (\mathbf{r}_i) = \left(\frac{4\pi}{2\ell + 1} \right)^{1/2} \mathbf{r}_i^\ell Y_\ell^m (\theta_i, \phi_i)$$

(Y_ℓ^m is the familiar spherical harmonic function). For $\ell = 0$, the leading term in the interaction energy is

$$U_0 = - \frac{1}{2} \left(1 - \frac{1}{\epsilon} \right) \frac{Q_0^2}{a} \tag{8-26}$$

TABLE 8-2
Solvation Energies (kJ/mole) for an Ethyl Fluoride Molecule in a Cavity of 114 $\mathring{A}^3$

			U_ℓ			
ϵ	U_1	U_2	U_3	U_4	U_5	Total U
3	2.28	0.422	0.067	0.021	0.017	2.80
8	3.36	0.619	0.063	0.029	0.025	4.15
78	4.07	0.748	0.134	0.038	0.033	5.024

Source: Ref. 46.

where Q_0 is the net charge of the solute molecule. For $\ell = 1$, the interaction energy is

$$U_1 = -\frac{1}{2}\left[\frac{2(\epsilon - 1)}{2\epsilon + 1}\right]\frac{\mu^2}{a^3} \qquad (8\text{-}36)$$

and so forth.

In a quantum-mechanical calculation based on this model, the solvation energy U depends on the wave function through the *electronic* contribution to the moments M_ℓ^m. Therefore, an additional term, depending on these electronic moments, occurs in the Hartree-Fock Hamiltonian. Table 8-2 shows the numerical results[46] for the different contributions U_ℓ for an ethyl fluoride molecule in a solvent "continuum" with dielectric constant ϵ. It is noticeable that the dipolar term U_1 alone contributes 80% of the total solvation energy.

A weakness of the method is the arbitrariness in the choice of the cavity radius a. Rivail and Rinaldi[46] use the approximate equation

$$\frac{a^3}{\alpha} \approx 3 \qquad (8\text{-}37)$$

which relates the radius to the molecular polarizability α, which they compute by a variational technique. Another problem arises from the inclusion in the Hartree-Fock matrix elements of that part of the electronic density lying outside the cavity radius—in contradiction with the assumptions of the model that the molecule be contained *within* the cavity. However, the contributions to (8-32) of the electronic density outside the cavity are found to be negligible.

8.7 Theories Using Discrete Solvent Molecules in a Model Form

We now turn to theories in which the solvent molecules are described as true discrete entities but in some simplified model form. These theories may be of the semicontinuum type if the distant bulk solvent is accounted for, or of the fully discrete type if the solvent description includes a (large) number of molecules.

In the latter category a powerful method has recently been described by Rossky, Karplus, and Rahman[48] for a peptide in water solution. All the necessary inter-molecular potential functions are provided, whereas the water molecules are de-scribed by the "ST2" model due to Stillinger and Rahman.[49,50] In this model each water molecule consists of four-point charges located within a single Lennard-Jones sphere. In the rigid ST2 model the four-point charges are fixed in a tetrahedrally oriented array around the central oxygen atom (Fig. 8-14). Two of these charges are positive ($-0.2357e$) and represent the hydrogen atoms; the other two are negative ($0.2357e$) and idealize the lone-pair orbital densities. The dipole moment of the model water molecule is 2.35 debye (experimental value 1.83 debye). The model can be made "flexible" by inclusion of OH bond stretching and HOH angle bending, while the position of the lone-pair charges "follows" in a reasonable manner that of the hydrogen atom charges.

The interaction between water molecules is governed by a refined potential function, which adds an electrostatic term to an ordinary Lennard-Jones 6-12 po-tential (exclusion repulsion proportional to the inverse twelfth power of the dis-tance, dispersion attraction proportional to the inverse sixth power of the dis-tance)[51]:

$$V_{OH_2 \leftrightarrow OH_2} = 4\epsilon_w\left[\left(\frac{\sigma_w}{r_{O_1O_2}}\right)^{12} - \left(\frac{\sigma_w}{r_{O_1O_2}}\right)^6\right] + \sum_{i,j=1}^{4} \frac{q_i^1 q_j^2}{r_{ij}} S(r_{O_1O_2}) \qquad (8\text{-}38)$$

where q_i^1 is a charge on the first water molecule and q_j^2 is a charge on the second molecule. The two oxygen atoms are O_1 and O_2. The parameters are

$$\epsilon_w = 0.317 \text{ kJ/mole} \qquad (8\text{-}39)$$
$$\sigma_w = 3.10 \text{ Å}$$

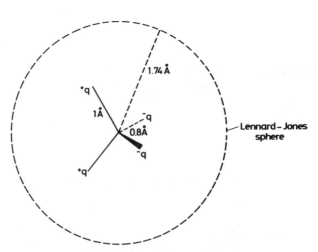

FIGURE 8-14. Rigid "ST2" model for water.

The minimum in the Lennard-Jones potential corresponds to an equilibrium inter-molecular oxygen-oxygen distance of $2^{1/6}\sigma_w = 3.48$ Å and a 1.74 Å radius for each individual molecular Lennard-Jones sphere. The function S is a "switching function" that modulates the electrostatic interaction in such a manner that the overall behavior mimics soft charged spheres at distances greater than $R_U = 3.1287$ Å and hard uncharged spheres at distances smaller than $R_L = 2.016$ Å:

$$S(r) = \begin{cases} 0 & 0 \leqslant r < R_L \\ \dfrac{(r - R_L)^2 (3R_U - R_L - 2r)}{(R_U - R_L)^3} & R_L \leqslant r \leqslant R_U \\ 1 & R_U \leqslant r \end{cases} \qquad (8\text{-}40)$$

Finally, the interactions between solute and solvent are described by a mixed Lennard-Jones–electrostatic potential

$$V_{\text{solute} \leftrightarrow \text{OH}_2} = \sum_{\substack{\text{all} \\ \text{solute} \\ \text{atoms } \lambda}} \left\{ 4\sqrt{\epsilon_w \epsilon_\lambda} \left[\left(\frac{\bar{\sigma}_\lambda}{r_{O\lambda}} \right)^{12} - \left(\frac{\bar{\sigma}_\lambda}{r_{O\lambda}} \right)^{6} \right] + \sum_{j=1}^{4} \frac{q_j q_\lambda}{r_{j\lambda}} \right\}$$

$$(8\text{-}41)$$

$$\bar{\sigma}_\lambda = \frac{\sigma_w + \sigma_\lambda}{2}$$

where q_λ represents the net charge on atom λ of the solute, and q_j represents the jth net charge of the water molecule. Typical numerical values of the Lennard-Jones parameters are shown in Table 8-3; the atomic charges q_λ vary with the nature of the solute.

The Rossky-Karplus-Rahman method also allows for the possibility of hydrogen bonding, for instance, between water molecules and a peptide N—H or C=O group. It suffices that the interaction potentials reproduce correctly experimental and theoretical estimates for the corresponding association energies. The relative

TABLE 8-3
Lennard-Jones Parameters[a]

Atom λ	ϵ_λ (kJ/mole)	σ_λ (Å)
C	0.376	3.208
H	0.0188	2.616
N	0.669	2.770
O	0.963	2.640

Source: Ref. 48
[a]The value σ is 1.78 times the radius of the Lennard-Jones sphere surrounding the atom.

TABLE 8-4
Relative Energies of Optimal Hydrogen Bonds[a]

	$-\epsilon$ (kJ/mole)
NH $\cdots$ OH$_2$	-25
NH $\cdots$ O$=$C	-34
OH$_2$ $\cdots$ OH$_2$	-28
C$=$O $\cdots$ H$_2$O	-31

Source: Ref. 48.
[a]Zero of energy: separated rigid species.

depths of different hydrogen bonds are shown in Table 8.4. The Lennard-Jones parameters and the charges q_λ are chosen to fit these depths as well as the correct energy behavior with geometric (angular) changes. With these potentials, the existence or nonexistence of hydrogen bonds depends on the calculated molecular configuration. The hydrogen bonds "are free to form or break in a continuous manner,"[48] and the existence of a given hydrogen bond can be inferred only from examination of the optimized geometry.

Earlier methods that used discrete solvent molecules in some model form were less sophisticated but already gave realistic results. For instance, a very simple model for water[52] uses fractional charges at the three atomic centers. Huron and Claverie,[53] on the other hand, studied various hydrocarbon solutes surrounded by six nitrobenzene molecules; the interaction energies were evaluated by using bond polarizabilities and bond charges on all molecules. The position of each solvent molecule relative to solute is allowed to vary through three angles.

Recently another rather sophisticated model of the semicontinuum type has been introduced.[54] It distinguishes four regions: (1) the solute region; (2) the strong field solvation shell region with mobile solvent molecules; (3) a bulk surface of "frozen" solvent molecules (at positions that correspond to the known structure of the bulk solvent); and (4) the continuum. In regions 2 and 3 the solvent molecules are represented as point dipoles attached to the centers of soft spheres. The contribution of the continuum to the solvation energy is estimated by taking the electrostatic energy of a sphere, with radius approximately that of the cavity enclosing regions 1–3, imbedded in a continuum with the bulk dielectric constant [see (8-26)].

The methods in this section are generally sufficiently simple to allow a *statistical exploration* of configurations, followed by a Boltzmann averaging, rather than a simple optimization of energy. Each configuration k is given a Boltzmann weighting factor:

$$p_k = \frac{e^{-E_k/RT}}{\displaystyle\sum_{\substack{\text{all} \\ \text{configurations } \ell}} e^{-E_\ell/RT}} \qquad (8\text{-}42)$$

and the mean value of the solute-solvent interaction energy becomes

$$\overline{U} = \sum_k p_k E_k \qquad (8\text{-}43)$$

8.8 Theories Using Discrete Solvent Molecules in an Exact Quantum-Mechanical Manner

The ultimate aim of a study of solvation is to acquire the capacity to include all the solvent molecules, at least many hundreds of them, with the solute in one single calculation. The level of quantum-mechanical calculation should be as accurate as possible, to avoid a low-quality treatment of the quantum-chemical problem as opposed to the high-quality treatment of the solvent. This approach, in which "all" the solvent molecules are included explicitly, is often called the *supermolecule* approach and has been pioneered by Alberte Pullman.[55,56] In practice, with the present computational facilities, only one or two solvent shells can be introduced. However, this is sufficient to reveal trends in solvent configurations and solvent surroundings for typical solutes.

The first stage in the approach is to start with a single solvent molecule. One explores the energy of interaction with the solute molecule for every possible reasonable configuration, approaching the solvent molecule from various directions, turning it around the solute, and rotating it about its local axes so as to find the best orientation. Not only is this equilibrium orientation found, together with the association energy, but the spanning of the hypersurface of interaction provides information on the "lability characteristics" of the solute-solvent binding. Pullman distinguishes three different types of lability: (1) *In situ lability* (lability for the solvent molecule to rotate about its local axes); (2) *Ad situm lability* (possibility of moving the solvent molecule in the neighborhood of the most stable position and away from it); and (3) *Extra situm lability* (energy characteristics of solvent motion in a region that is not the stable site). These labilities can be illustrated (Fig. 8-15)

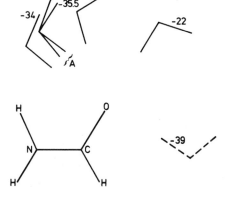

FIGURE 8-15. Various types of lability (A, *In situ;* B, *Ad situm;* C, *Extra situm*) for a solvent water molecule in the near-optimal binding site of formamide (energies in kJ/mole).[55] The optimal binding site is shown in dotted lines. Reprinted with permission from A. Pullman, p. 149 of *The New World of Quantum Chemistry,* edited by B. Pullman and R. G. Parr. Copyright 1976 by D. Reidel Publishing Company.

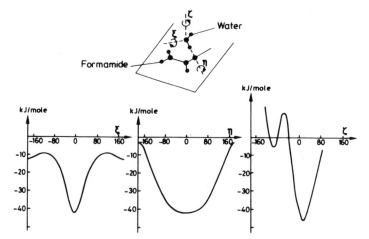

FIGURE 8-16.   Energy variation *in situ* for water in near-optimal (lateral) hydration site of formamide.[55,57] Reprinted with permission from A. Pullman, p. 149 of *The New World of Quantum Chemistry*, edited by B. Pullman and R. G. Parr. Copyright 1976 by D. Reidel Publishing Company

for one of the two optimal hydration sites of formamide. The energy variations for rotation *in situ* are shown in Fig. 8-16.[55,57]

The second stage is the buildup of the first solvation shell. The number of molecules that are directly bound to the solute is sometimes difficult to determine. An additional solvent molecule may have a significant binding energy with the solute, but the critical quantity is the total energy (solute plus all solvent molecules) relative to the molecules all at infinite distance from each other. If the repulsive energy with the preexisting solvent shell is too large, the additional solvent molecule will not "hold on" to this shell. First hydration shells have been obtained in this manner for a large number of solutes, such as formamide, urea (Fig. 8-17), formic acid, methanol, and so on. Examination of the preferential solvent sites in these polyhydrated shells reveal characteristic positions for the water molecules. A given chemical group always tends to be surrounded in the same manner. For

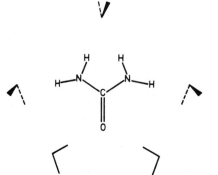

FIGURE 8-17.   The first hydration shell of urea.[56] Reprinted with permission from A. Pullman, p. 1 of *Quantum Theory of Chemical Reactions*, Vol. II, edited by R. Daudel, A. Pullman, L. Salem, and A. Veillard. Copyright 1981 by D. Reidel Publishing Company.

instance, a carbonyl group tends to have two water molecules hydrogen–bonded to its two in-plane sp^2 lone pairs (Fig. 8-17). An isolated amino group has two solvent water molecules, one near each amino hydrogen atom, with the OH_2 plane roughly perpendicular to the NH_2 plane and either the sp^2-σ oxygen lone pair or $p\pi$ oxygen lone pair—or sometimes the traditional sp^3 tetrahedral combination of the two (case of ethylamine)—pointing toward the the amino hydrogen. In urea both types of binding occur (σ lone pair for the central hydrogens of urea and p lone pair for the external hydrogens; see Fig. 8-17).

The final stage consists in computing properties of the solvent-solute super-molecule—activation energies and conformational barriers—in its optimal arrangement. As mentioned previously, this arrangement may itself depend on the reaction coordinate to be examined.

Theories that use discrete solvent molecules in an exact quantum-mechanical manner can reveal fine details that other theories are unable to provide. For instance, adenine has four water molecules in its first solvation shell and thymine, three. But the adenine-thymine base pair has only six solvent molecules; base pairing is accompanied by extrusion of one water molecule.[58] Another example concerns the structure of water around nonpolar methyl groups. Calculations on the dimethyl phosphate anion[59] show that the water molecules bind only weakly to the methyl groups but hydrogen bond strongly among themselves to form an ordered network similar to the "icebergs" proposed long ago by Frank and Evans.[60] This confirms calculations by the ST2 model (Section 8.7) on a dipeptide in water that show[61] that none of the hydrogen atoms nor lone-pair orbitals of the water molecules point at the nonpolar groups. Such solvation simply maximizes the number of favorable water-water interactions.

8.9 Chemical Reactions Induced by Solvent Motion: Electron Transfer

We showed in Section 5.8 that an electron jump or electron transfer occurs when two approximate potential surfaces, one covalent and the other ionic, intersect (Fig. 5-15). This intersection may or may not be washed out when the fully correct surfaces are drawn out; the avoided or real intersection signals the electron transfer region.

The crossing of covalent and ionic surfaces may be induced, as in NaCl, simply by a change in distance between partners A and B. The rapidly changing $1/R$-like Coulomb attraction between the ionic partners A^+ and B^- causes a rapid displacement of the ionic surface (A^+, B^-) relative to the covalent surface (A,B). This feature is illustrated in Fig. 8-18a for the interaction between ammonia and cyanoethylene in excited states of the combined system.[62,63] The "charge-transfer" excited state of high energy represented by (ammonia$^+$, cyanoethylene$^-$) decreases rapidly in energy as the molecules approach. It soon crosses the "locally excited" state (ammonia, cyanoethylene in its $^1\pi,\pi^*$ state) whose energy is relatively insensitive to the intermolecular distance. On the lower surface, electron transfer occurs from ammonia to cyanoethylene, whereas the opposite occurs on the higher surface.

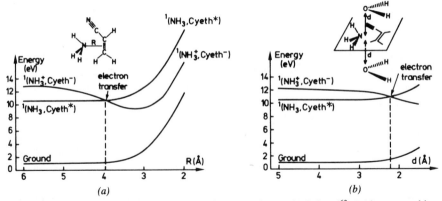

FIGURE 8-18.   Different manners in which electron transfer can be induced[62]: *(a)* by approaching partners; *(b)* by moving solvent molecules (*R* fixed at 5 Å).

However, the same intersection, and concomitant electron transfer, can be obtained by keeping the partners at fixed distance and simply reorienting the solvent relative to them. It is convenient to place two water molecules above and below the symmetry plane of the system in such a manner as to stabilize the would-be charge transfer pair (the oxygen atoms thus point at the positive-to-be nitrogen atoms). As the water molecules approach the reaction plane, the ionic surface is again strongly stabilized and intersects the covalent surface (Fig. 8-18*b*)[62]; electron transfer is again induced.

Figure 8-19 shows a three-dimensional mapping of the two surfaces[63], in which the region of photochemical electron transfer appears clearly. The most striking feature of this calculation is the proof that electron transfer is essentially a *reaction of solvent molecules,* with some assistance from the interacting partners that must approach at a reasonable distance. The electron jump from one partner to the other is never spontaneous; it requires appropriate *prior motions of the surrounding solution.* It also obeys geometric requirements*: (1) a nonzero overlap between the clouds of the donating molecular orbital and of the accepting molecular orbital (hence the necessity that the partners be not too far apart); (2) a favorable energy relationship between these two orbitals (Chapter 6).

Quantitative theories of electron transfer have been developed by Marcus[64–66] and by the school of Levich.[67,68] These theories and others[69] are based on a picture of two nearly intersecting surfaces on a potential energy diagram, where the reaction coordinate involves some unspecified mixture of the nuclear coordinates of the reactants with those of the solvent (Fig. 8-20). Typical reactant donor-acceptor pairs might be $Fe(CN)_6^{3-}$, $Fe(CN)_6^{4-}$, or $HN(CH_3)_2$, CH_3COOH.

Important parameters in the theory are the free energy of activation $\Delta G^{\ddagger}$ and the Marcus parameter λ, defined as the energy required to transfer the electron

*S. F. Nelsen points out to the author that, generally, electron transfer becomes exceedingly slow when large geometry changes are involved.

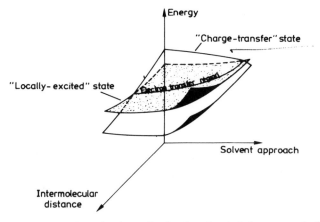

FIGURE 8-19.   Three-dimensional mapping for photochemical electron transfer.[62,63]

from reactant at its equilibrium geometry to the product in its own equilibrium geometry (hence λ in Figure 8-20 is measured up to an energy which corresponds to the intersection of reactant curve with product geometry). Marcus[65] shows that

$$\lambda = (\Delta e)^2 \left(\frac{1}{2a_1} + \frac{1}{2a_2} - \frac{1}{r} \right) \left(\frac{1}{n^2} - \frac{1}{\epsilon} \right) \tag{8-44}$$

where Δe is the charge transferred, a_1 and a_2 are the ionic radii of the ions involved (including their coordination shells), r is their distance which is roughly $a_1 + a_2$, and n and ϵ are the refractive index and the dielectric constant, respectively. Marcus relates $\Delta G^{\ddagger}$ to λ and to the work required to bring the reactants together from infinity to their distance r at the prevailing salt concentration. The theory can then be tested by comparing calculated activation energies with experimental ones since

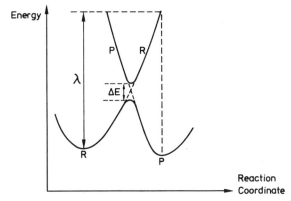

FIGURE 8-20.   Profile of nearly intersecting surfaces for an electron transfer reaction.[65] Curve R denotes (donor + acceptor); curve P denotes (donor$^{\oplus}$ + acceptor$^{\ominus}$).

all the quantities in (8-44) are known.* The theory is extremely useful in electro-chemistry since it allows predictions of rate constants for electron transfer in redox reactions.

It should be noted that the splitting ΔE between the two surfaces (reactants R, products P) at the electron transfer point is:

$$\Delta E = 2H_{RP} \tag{8-45}$$

[see (5-39) and (5-52)], where H is the full Hamiltonian that allows for mixing of covalent and ionic terms. The matrix element H_{RP} between an ionic state of simple wave function

$$R \equiv A(1) \, A(2) \tag{8-46}$$

and a covalent state

$$P \equiv A(1) \, B(2) + B(1) \, A(2) \tag{8-47}$$

is proportional to

$$H_{RP} = \int A(1) \, A(2) \, H \, [A(1) \, B(2) + B(1) \, A(2)] d\tau_1 \, d\tau_2 \tag{8-48}$$

The size of H_{RP} clearly depends on the size of the overlap integral

$$S_{AB} = \int AB \, d\tau \tag{8-49}$$

between the orbitals of the reacting partners. The larger this overlap, the larger H_{RP} and the splitting ΔE, and the smaller the barrier for electron transfer.[70] This con-firms our statement that overlap is an important geometric constraint in electron transfer.

8.10   Tight and Solvent-Separated Ion Pairs

The existence of two different types of ion pair in solution, with different solvent-dependent degrees of dissociation, was postulated by Winstein[71] in his study of acetolysis of benzenesulfonates in presence of lithium perchlorate. A special salt effect alters the rate of production of carbenium ions and is attributed to the ex-istence of two types of ion pairs, only one of which, the "external" or "solvent-separated" ion pair, is affected by the salt. The full Winstein scheme is[72]

---

*The author is grateful to C. Amatore of the University of Paris for a discussion of Marcus' theory.

$$
\begin{array}{ccc}
\text{``Internal''} & \text{``Solvent-} & \\
\text{or ``tight''} & \text{separated''} & \text{``Free ions''}
\end{array}
$$

$$
(R^{\delta+}\!\!-\!X^{\delta-})_{solv} \rightleftarrows (R^+\,X^-)_{solv} \rightleftarrows (R^+/X^-)_{solv} \rightleftarrows (R^+)_{solv} + (X^-)_{solv}
$$

$$
\begin{array}{ccc}
\downarrow & \downarrow & \downarrow \\
\text{products} & \text{products} & \text{products}
\end{array}
$$

$$(8\text{-}50)$$

The potential-energy surface for ion-pair dissociation, which could eventually reveal the two types of ion pair, has been a recent challenge for theoreticians. One of the first calculations is due to Simonetta[73] for CH_3F surrounded by a model solvent of 11 water molecules. The full "supermolecule" approach (Section 8.8) was used, but with the CNDO/2 method of evaluating energies (Section 1.9). The potential energy curve for total energy-versus-C—F distance shows three minima (Fig. 8-21). These minima are interpreted respectively as (1) a bonded CH_3F molecule with slightly stretched bond length (1.385 versus 1.344 Å at equilibrium), with increased charge on fluorine and carbon, indicating a tendency toward a "tight" ion pair, (2) a highly stretched molecule with two solvent molecules on either side of, but very close to, the CF bond region (Fig. 8-21), and (3) a truly *solvent-separated* situation in which two water molecules have penetrated completely between carbon and fluorine. The two ions are now in two different, but adjacent, solvent cages. Recently Bigot and the present author have developed a model[74] with two hard-sphere point charges surrounded by a number of hard-sphere solvent dipole molecules. The actual number of such molecules in the first solvent shell—defined as those in direct contact with the ions—is considered to be variable

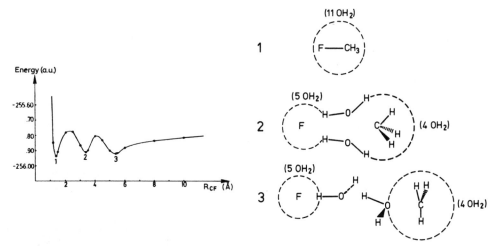

FIGURE 8-21. Total energy versus C—F distance for dissociation of CH_3F in the presence of 11 water molecules[73] and schematic representation of minima 1, 2, and 3. Magnitude of the energy wells (>100 kJ/mole) is exaggerated by the method of calculation. Reprinted with permission from P. Cremaschi, A. Gamba, and M. Simonetta *J. Chem. Soc., Perkin II* 162 (1977). Copyright 1977 The Chemical Society.

and optimized to lower the energy. A two-dimensional electrostatic calculation gives an optimum number of seven solvent molecules (radius 1.5 Å) around each ion (radius 2.2 Å). The energy as a function of cation-anion distance then shows four separate minima (Fig. 8-22). Their interpretation is akin to Simonetta's minima: first a "tight" ion pair (12); next a "slightly stretched" ion pair (9) with two solvent molecules close to the internuclear axis; a truly *solvent-separated* ion pair (7) with one solvent molecule between the ions; and a second form of solvent-separated ion pair (4) with two pairs of solvent molecules sideways between the ions. Structures 7 and 4 bear great resemblance to Simonetta's structure 3. The final structure 1 in the present scheme corresponds to the free ions. The energy dips all have have a magnitude of approximately 30 kJ/mole.

Hence both models, despite their relatively crude nature, lead to the "tight" ion pair and the "solvent-separated" ion pair postulated by Winstein. Only one or two solvent molecules separate the ions in the latter. Particularly interesting are the stable systems 2[73] and 9,[74] which are halfway between "tight" and "solvent-separated" and where two solvent molecules have started "prying" the ions apart. They might be called *"tentatively"* separated ion pairs.

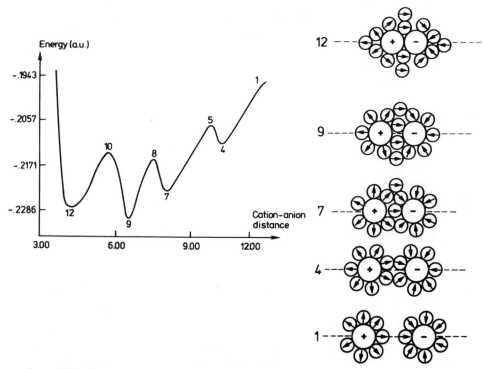

FIGURE 8-22.   Energy versus cation-anion distance in two-dimensional electrostatic model of two hard-sphere ions surrounded by 30 hard-sphere dipoles.[74] Schematic representation of minima is shown on the right.

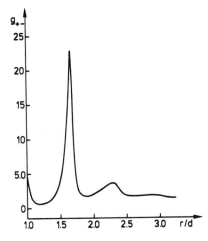

FIGURE 8-23.   Like-unlike ion-pair distribution function g_{+-}.[75] The interionic distance is r, and d is the common diameter of positive and negative ions; the ionic concentration corresponds to an 0.05 molarity. The other variables are 2.5 for the reduced dipole of the solvent molecules $[\mu^2/(kTd_\mu^3)]$, 136 for the reduced charge of the ions $[q^2/(kTd)]$, and 0.6 for the reduced dipole number density $(\rho_\mu d_\mu^3)$. We have $d_\mu = 0.68d$. Reprinted with permission from D. Levesque, J. J. Weis, and G. N. Patey, *J. Chem. Phys.* **72,** 1887 (1980). Copyright 1980 by American Institute of Physics.

Much more sophisticated methods, based on statistical mechanics, can yield ion–ion pair distribution functions. A typical result[75] for the like-unlike ion-pair distribution function, is shown in Fig. 8-23. The peak at an interionic distance $r = 1.68$ (d = diameter of the ions, whereas the diameter d_μ of the dipolar solvent molecules is $0.68\ d$) is due to pairs separated by a single solvent molecule. It is higher than the peak at $r = d$ for ions in contact. Hence for this particular system, there are more solvent-separated ion pairs than tight ion pairs. Naturally, the relative height of the solvent-separated ion pair peak decreases with increasing ion concentration, indicating a lesser probability of finding this species.

References

1.   A. J. Parker, *Quart. Rev.* **16,** 163 (1962).
2.   I. Roberts and G. E. Kimball, *J. Am. Chem. Soc.* **59,** 947 (1937).
3.   B. Tchoubar, *Bull. Soc. Chim. Fr.* 2069 (1964).
4.   E. D. Hughes and C. K. Ingold, *J. Chem. Soc.* 244 (1935); *Transact. Faraday Soc.* **37,** 603, 657 (1941).
5.   C. Reichardt, *Solvent Effects in Organic Chemistry* (Verlag Chemie, Weinheim, 1979), p. 85.
6.   H. Margenau and W. W. Watson, *Rev. Mod. Phys.* **8,** 22 (1936).
7.   N. S. Bayliss, *J. Chem. Phys.* **18,** 292 (1950).
8.   H. C. Longuet-Higgins and J. A. Pople, *J. Chem. Phys.* **27,** 192 (1957).
9.   E. G. McRae, *J. Phys. Chem.* **61,** 562 (1957).
10.   W. Liptay, *Angew. Chem. Internatl. Ed.* **8,** 177 (1969).
11.   N. Mataga and T. Kubota, *Molecular Interactions and Electronic Spectra* (Dekker, New York, 1970), Chapter 8.
12.   E. Lippert, *Acc. Chem. Res.* **3,** 74 (1970).
13.   A. T. Amos and B. L. Burrows, *Adv. Quant. Chem.* **7,** 289 (1973).
14.   M. Lamotte, D.Sc. thesis, University of Bordeaux I, France, 1973.
15.   L. Onsager, *J. Am. Chem. Soc.* **58,** 1486 (1936).
16.   J. Franck, *Transact. Faraday Soc.* **21,** 536 (1925).
17.   E. U. Condon, *Phys. Rev.* **32,** 858 (1928).
18:   Ref. 5, p. 197.

19. S. Iwata and K. Morokuma, *J. Am. Chem. Soc.* **97**, 966 (1975).
20. P. Haberfield, *J. Am. Chem. Soc.* **96**, 6526 (1974); P. Haberfield, M. S. Lux, and D. Rosen, ibid. **99**, 6828 (1977).
21. J. I. Brauman and L. K. Blair, *J. Am. Chem. Soc.* **92**, 5986 (1970); ibid. **93**, 3911 (1971).
22. R. F. Hudson, O. Eisenstein, and N. T. Anh, *Tetrahedron* **31**, 751 (1975).
23. W. L. Jorgensen and L. Salem, The Organic Chemist's Book of Orbitals (Academic, New York, 1974).
24. J. A. Pople and M. Gordon, *J. Am. Chem. Soc.* **89**, 4253 (1967); W. J. Hehre and J. A. Pople, ibid. **92**, 219 (1970).
25. A. Pross and L. Radom, *J. Am. Chem. Soc.* **100**, 6572 (1978).
26. D. J. DeFrees, J. E. Bartness, J. K. Kim, R. T. McIver, and W. J. Hehre, *J. Am. Chem. Soc.* **99**, 6451 (1977).
27. C. Minot and N. T. Anh, *Tetrahedron Lett.* 3905 (1975).
28. W. L. Jorgensen, *J. Am. Chem. Soc.* **100**, 1049 (1978).
29. F. S. Dainton, *Chem. Soc. Rev.* **4**, 323 (1975).
30. M. S. Matheson in *Physical Chemistry*, Vol. VII, *Reactions in Condensed Phases*, edited by H. Eyring, D. Henderson, and W. Jost (Academic, New York, 1975), Chapter 10.
31. W. Kauzmann, *Quantum Chemistry, An Introduction* (Academic, New York, 1957), p. 187.
32. M. D. Newton, *J. Phys. Chem.* **79**, 2795 (1975).
33. M. Natori and T. Watanabe, *J. Phys. Soc. Jap.*, **21**, 1573 (1966); M. Natori, ibid. **24**, 913 (1968); ibid. **27**, 1309 (1969).
34. M. Weissmann and N. V. Cohan, *Chem. Phys. Lett.* **7**, 445 (1970).
35. D. A. Copeland, N. R. Kestner, and J. Jortner, *J. Chem. Phys.* **53**, 1189 (1970).
36. K. Fueki, D. F. Feng, and L. Kevan, *J. Am. Chem. Soc.* **95**, 1398 (1973).
37. J. W. Moskowitz, M. Boring, and J. H. Wood, *J. Chem. Phys.* **62**, 2254 (1975).
38. C. A. Coulson, *Electricity* (Oliver and Boyd, London, 1956), p. 66.
39. G. Klopman, *Chem. Phys. Lett.* **1**, 200 (1967).
40. M. Born, *Z. Physik* **1**, 45 (1920).
41. R. Constanciel and O. Tapia, *Theoret. Chim. Acta (Berl.)* **48**, 75 (1978).
42. R. Constanciel, *Theoret. Chim. Acta (Berl.)* **54**, 123 (1980).
43. R. Constanciel and O. Tapia, unpublished results.
44. J. E. Sanhueza, O. Tapia, and C. Lamborelle, *TN-621, Quantum Chemistry Group*, Uppsala University, Sweden.
45. J. Hylton, R. E. Christoffersen, and G. G. Hall, *Chem. Phys. Lett.* **26**, 501 (1974); J. Hylton McCreery, R. E. Christoffersen, and G. G. Hall, *J. Am. Chem. Soc.* **98**, 7191, 7198 (1976).
46. J. L. Rivail and D. Rinaldi, *Chem. Phys.* **18**, 233 (1976).
47. J. G. Kirkwood, *J. Chem. Phys.* **2**, 351 (1934); J. G. Kirkwood and F. H. Westheimer, ibid. **6**, 506 (1938).
48. P. J. Rossky, M. Karplus, and A. Rahman, *Biopolymers* **18**, 825 (1979).
49. F. H. Stillinger and A. Rahman, *J. Chem. Phys.* **60**, 1545 (1974).
50. A. Rahman and F. H. Stillinger, *J. Am. Chem. Soc.* **95**, 7943 (1973).
51. J. O. Hirschfelder, C. F. Curtiss, and R. B. Bird, *Molecular Theory of Gases and Liquids* (Wiley, New York, 1954).
52. J. O. Noell and K. Morokuma, *Chem. Phys. Lett.* **36**, 465 (1975).
53. M.-J. Huron and P. Claverie, *Chem. Phys. Lett.* **9**, 194 (1971).
54. A. Warshel, *Chem. Phys. Lett.* **55**, 454 (1978); *J. Phys. Chem.* **83**, 1640 (1979).
55. A. Pullman in *The New World of Quantum Chemistry, Proceedings of the Second International Congress of Quantum Chemistry*, edited by B. Pullmann and R. Parr (Reidel, Dordrecht, 1976), p. 149.
56. A. Pullman in *Quantum Theory of Chemical Reactions*, Vol. II, edited by R. Daudel, A. Pullman, L. Salem, and A. Veillard (Reidel, Dordrecht, 1981), p. 1.
57. G. Alagona, A. Pullman, E. Scrocco, and J. Tomasi, *Internatl. J. Peptide Protein Res.* **5**, 251 (1973).
58. B. Pullman, S. Miertus, and D. Perahia, *Theoret. Chem. Acta (Berl.)* **50**, 317 (1979).

59. J. Langlet, P. Claverie, B, Pullman, and D. Piazzola, *Internatl. J. Quant. Chem., Quantum Biol. Symp.* **6,** 409 (1979).
60. H. S. Frank and M. W. Evans, *J. Chem. Phys.* **13,** 507 (1945).
61. P. J. Rossky and M. Karplus, *J. Am. Chem. Soc.* **101,** 1913 (1979).
62. G. Ramunni and L. Salem, *Z. Physik. Chemie Neue Folge* **101,** 123 (1976).
63. L. Salem, *Science* **191,** 822 (1976).
64. R. Marcus, *J. Chem. Phys.* **24.** 966 (1956).
65. R. Marcus, *Disc. Faraday Soc.* **29,** 21 (1960).
66. R. Marcus, *J. Chem. Phys.* **43,** 679 (1965).
67. V. G. Levich and R. R. Dogonadze, *Dokl. Akad. Nauk SSSR* **133,** 158 (1960).
68. E. D. German, R. R. Dogonadze, A. M. Kuznetsov, V. G. Levich, and Y. I. Kharkats, *J. Res. Inst. Catalysis (Hokkaido Univ.)* **19,** 99, 115 (1971).
69. N. S. Hush, *Transact. Faraday Soc.* **57,** 557 (1961).
70. D. Roux, Thèse de 3e Cycle, Université de Paris-Sud, Orsay, France, 1980, p. 107.
71. S. Winstein, E. Clippinger, A. H. Fainberg, and G. C. Robinson, *J. Am. Chem. Soc.* **76,** 2597 (1954).
72. Ref. 5, p. 35.
73. P. Cremaschi, A. Gamba, and M. Simonetta, *J. Chem. Soc., Perkin II*, 162 (1977).
74. B. Bigot and L. Salem, unpublished results.
75. D. Levesque, J. J. Weis, and G. N. Patey, *J. Chem. Phys.* **72,** 1887 (1980).

Author Index

Page numbers in roman indicate citations within chapters; page numbers *in italics* indicate reference lists at the end of each chapter.

Subject Index

BELMONT COLLEGE LIBRARY